Development Aid and Adaptation to Climate Change in Developing Countries

Carola Betzold • Florian Weiler

Development Aid and Adaptation to Climate Change in Developing Countries

palgrave
macmillan

Carola Betzold
Institute of Political Science
University of Göttingen
Göttingen, Germany

Florian Weiler
The School of Politics
and International Relations
University of Kent
Canterbury, Kent
UK

ISBN 978-3-319-64509-4 ISBN 978-3-319-64510-0 (eBook)
https://doi.org/10.1007/978-3-319-64510-0

Library of Congress Control Number: 2017954953

Cover illustration: Artpilot

Printed on acid-free paper

This Palgrave Macmillan imprint is published by Springer Nature
The registered company is Springer International Publishing AG
The registered company address is: Gewerbestrasse 11, 6330 Cham, Switzerland

Preface

Developed countries have a moral and a legal obligation to assist developing countries that are 'particularly' vulnerable to the adverse effects of climate change to deal with these effects. Financial flows from the Global North for adaptation measures in the Global South are thus on the rise. To what extent do these financial flows reach those most in need, those most vulnerable to climate change and least able to adapt? And, just as importantly, to what extent does the support from the Global North help to reduce vulnerability and increase resilience among recipients?

In this book we focus on the first question: to what extent does adaptation finance reach vulnerable countries? This speaks to the larger question of aid allocation: how do donors distribute their adaptation aid? What factors and criteria do they consider when making allocation decisions? And how does the allocation of aid for adaptation differ from the allocation of development aid in general? We started to become interested in these questions as we observed that climate finance and specifically adaptation finance is increasing globally, but that not all vulnerable countries such as the small island developing states (SIDS) benefit to the same extent from this new form of finance. While we were first interested in explaining variation across SIDS only, we quickly extended our line of research to include all developing countries eligible to receive development assistance.

We started to examine adaptation aid allocation by assessing data from the OECD and presented the first results at the International Studies Association Annual Conference in 2014, and there met with Christina Brian from Palgrave Macmillan. She encouraged us to extend our research and turn the conference paper into a book-length project. Without this encouragement, we would not have written this book, and so we would like to thank Christina first and foremost. We would also like to extend our gratitude to the entire editing and production team for their support.

While statistical analyses of aid data served as the starting point of our research project, we noted that numbers can only tell us so much. We hence extended our first attempt at understanding how adaptation aid is distributed through in-depth qualitative case studies. These built mainly on semi-structured interviews with practitioners and other stakeholders and observers in the aid sector in Germany, Sweden, and the United Kingdom, our case study countries. We are very grateful to all our interview partners for taking the time to speak to us and sharing their experiences and perspectives. Funding from the Fritz Thyssen Foundation made it possible for us to travel to Germany, Sweden, and the United Kingdom to conduct interviews face-to-face.

We would also like to thank the many colleagues who helped develop our research and this book. Thanks in particular to Anaïd Flesken for helpful comments on previous versions of the book. Participants at the workshop 'Climate Finance: Taking Stock, Future Directions for Policy and Research' held in Lund in April 2015 provided useful comments on parts of the project, as did Katja Michaelowa, Axel Michaelowa, Paula Castro, and participants at the Climate and Environment Workshop held in Zurich in January 2015. Matthew Dornan contributed to our larger research project and provided valuable input on the aggregate analysis of all donors. Tobias Jakobi helped with some technical questions. All errors remain of course our own.

Göttingen, Germany
Basel, Switzerland

Carola Betzold
Florian Weiler

Fritz Thyssen Stiftung
für Wissenschaftsförderung

Contents

1 Introduction 1

1.1 Definition of Key Concepts 3

1.1.1 Adaptation to Climate Change 4

1.1.2 Vulnerability 4

1.2 The Cost of Climate Change Adaptation 7

1.3 Financing Climate Change Adaptation 9

1.4 Overview of the Book and Contributions 11

References 15

2 The History and Political Economy of Adaptation Aid 21

2.1 History of Adaptation and Adaptation Finance in Climate Change Negotiations 22

2.1.1 Adaptation at the Sidelines: The Early Years of Climate Change Negotiations 22

2.1.2 Adaptation to the Fore: Adaptation Finance from 2001 to 2009 25

2.1.3 Finance Breakthrough: Adaptation Finance Since 2009 26

2.2 Adaptation Finance or Adaptation Aid? The Question of Additionality 28

2.3 *Aid Allocation* 31
2.3.1 *Recipient Need vs Donor Interests* 32
2.3.2 *Recipient Merit* 35
2.3.3 *Other Factors* 36
2.4 *Allocating Adaptation Aid* 37
2.4.1 *Vulnerability as Recipient Need* 38
2.4.2 *Measuring Vulnerability* 40
2.4.3 *Does Vulnerability Matter?* 42
2.5 *Summary and Our Expectations* 44
References 46

3 Setting the Stage: A Mixed Methods Research Design 57
3.1 *Our Overall Research Design: Mixed Methods* 58
3.2 *Quantitative Design* 60
3.2.1 *Our Dependent Variable: Adaptation Aid* 62
3.2.2 *Physical Vulnerability* 66
3.2.3 *Adaptive Capacity and Recipient Merit* 72
3.2.4 *Donor Interests* 76
3.2.5 *Control Variables* 79
3.2.6 *Modelling Strategy* 80
3.3 *Qualitative Design* 82
3.3.1 *Semi-structured Interviews* 83
3.3.2 *Policy Documents* 84
3.3.3 *Analysing Our Interview and Document Data* 86
References 90

4 Descriptive Analysis: Adaptation Aid Flows in the OECD CRS 99
4.1 *Overview of Adaptation Aid Flows* 100
4.1.1 *Adaptation Aid Compared to Total Aid* 100
4.1.2 *Commitments Compared to Disbursements* 102
4.2 *Who Gives Adaptation Aid?* 104
4.2.1 *Bilateral Compared to Multilateral Flows* 104
4.2.2 *Bilateral Donors in Comparison: Total Adaptation Aid* 105

4.2.3 *Bilateral Donors in Comparison: Per Capita and Percent* 107
4.3 *Who Receives Adaptation Aid?* 109
4.3.1 *Type of Recipient* 110
4.3.2 *Bilateral Versus Multilateral Flows* 111
4.3.3 *Recipients in Comparison* 113
References 118

5 Quantitative Analysis: Who Gives Adaptation Aid to Whom? 121
5.1 *Adaptation Aid Allocation Across All Donors* 123
5.1.1 *Physical Exposure and Sensitivity to Climate Change Impacts* 130
5.1.2 *Adaptive Capacity and Good Governance* 137
5.1.3 *Donor Interests* 149
5.1.4 *Control Variables* 155
5.2 *Adaptation Aid Allocation in Germany, Sweden, and the UK* 156
5.2.1 *Germany* 157
5.2.2 *Sweden* 163
5.2.3 *United Kingdom* 165
References 168

6 Qualitative Analysis: Adaptation Aid in Germany, Sweden, and the United Kingdom 171
6.1 *Climate Change, Adaptation, and Development Cooperation* 172
6.1.1 *Climate Change as a Priority Area* 173
6.1.2 *Adaptation or Development?* 175
6.1.3 *Adaptation or Mitigation?* 177
6.2 *The Aid Allocation Process* 178
6.2.1 *Recipient Need or Vulnerability to Climate Change* 180
6.2.2 *Recipient Merit* 183
6.2.3 *Donor Interests* 186
6.2.4 *Additional Factors* 188

6.3 Issues and Challenges 190
6.3.1 Additionality 190
6.3.2 Reporting and Accounting 192
6.3.3 Level of Funding 194
References 197

7 Conclusion 201
7.1 Overview 202
7.2 Results 205
7.3 Implications 209
References 214

Appendix 217

Index 219

List of Abbreviations and Acronyms

BMZ	Bundesministerium für wirtschaftliche Zusammenarbeit und Entwicklung (German Federal Ministry for Economic Cooperation and Development)
BRICS	Brazil, Russia, India, China, and South Africa
CDKN	Climate and Development Knowledge Network
COP	Conference of the Parties
CRI	Climate Risk Index
CRS	Creditor Reporting System
DAC	Development Assistance Committee
DEval	German Institute for Development Evaluation
DfID	Department for International Development
EVI	Environmental Vulnerability Index
G7	Group of Seven (Canada, France, Germany, Italy, Japan, the United Kingdom, and the United States)
GDP	Gross domestic product
GIZ	Deutsche Gesellschaft für International Zusammenarbeit (German Society for International Cooperation)
GNI	Gross national income
IPCC	Intergovernmental Panel on Climate Change
KfW	Kreditanstalt für Wiederaufbau
LDCF	Least Developed Country Fund
LDCs	Least developed countries
NAPAs	National Adaptation Programmes of Action
ND-GAIN	Notre Dame Global Adaptation Index
NGO	Non-governmental organisation
ODA	Official development assistance

OECD	Organisation for Economic Co-operation and Development
PPP	Purchasing power parity
SCCF	Special Climate Change Fund
Sida	Swedish International Development Cooperation Agency
SIDS	Small island developing states
SOPAC	South Pacific Applied Geoscience Commission
SVCCI	Structural Vulnerability to Climate Change Index
UAE	United Arab Emirates
UK	United Kingdom
UN	United Nations
UNDP	United Nations Development Programme
UNFCCC	United Nations Framework Convention on Climate Change
USA	United States of America
WGIs	Worldwide Governance Indicators

List of Figures

Fig. 1.1	Vulnerability and its components	5
Fig. 4.1	Total aid flows by year and adaptation objective	101
Fig. 4.2	Adaptation aid disbursements compared to commitments over time (**a**); disbursements in percent of commitments for selected donors (**b**)	103
Fig. 4.3	Adaptation aid flows by donor type	104
Fig. 4.4	Total adaptation aid for selected donors	106
Fig. 4.5	Adaptation aid for selected donors; per capita (**a**), and in percent of all development aid (**b**)	108
Fig. 4.6	Adaptation aid by recipient type; overall (**a**), and for selected donors (**b**)	110
Fig. 4.7	Share of adaptation aid provided by bilateral donors	112
Fig. 4.8	Distribution of principal adaptation aid	114
Fig. 4.9	Distribution of significant adaptation aid	115
Fig. 5.1	Correlation matrix of all dependent and independent variables	124
Fig. 5.2	Effects of the physical exposure variables for the full principal aid models of the selection stage (above) and the allocation stage (below), including 95% confidence intervals	131
Fig. 5.3	Effects of the adaptive capacity and/or recipient merit variables for the full principal aid models of the selection stage (above) and the allocation stage (below), including 95% confidence intervals	139
Fig. 5.4	Effects of the vulnerability country dummies for the full principal aid models of the selection stage (panel (**a**)) and the allocation stage (panel (**b**)), including 95% confidence intervals	144

Fig. 5.5 Effects of the donor interest variables for the full principal adaptation aid models of the selection stage (above) and the allocation stage (below), including 95% confidence intervals 151

Fig. 7.1 All substantive effects for the full model for both adaptation aid allocation stages. (**a**) Selection stage. (**b**) Allocation stage 206

List of Tables

Table 3.1	Summary of all variables used in the statistical analysis	83
Table 3.2	List of interviews conducted in Sweden, Germany, and the UK	85
Table 3.3	Overview of the policy documents used in the qualitative analysis	87
Table 5.1	Principal adaptation aid, selection stage, all countries	126
Table 5.2	Principal adaptation aid, allocation stage, all countries	127
Table 5.3	Principal and discounted significant adaptation aid, selection stage, all countries	128
Table 5.4	Principal and discounted significant adaptation aid, allocation stage, all countries	129
Table 5.5	Selection and allocation stage for Germany, Sweden, and the UK, principal adaptation aid only	158
Table 5.6	Selection and allocation stage for Germany, Sweden, and the UK, principal and (discounted) significant adaptation aid	160
Table A.1	Summary statistics for all variables used in the quantitative analysis	217
Table A.2	Questions guiding our semi-structured interviews	218

CHAPTER 1

Introduction

Open the newspaper on any given day and you will likely find reports on devastating weather extremes somewhere on the globe. For instance, in late September/early October 2016, hurricane Matthew made headlines. In Haïti, it left more than 500 people dead, tens of thousands homeless, and hundreds of thousands in need of food assistance (OCHA 2016). Economic damage across the hurricane's path through Haïti, Cuba, the Bahamas, and the USA's south-eastern coast is estimated at over $8 billion (AIR Worldwide 2016). In March 2015, tropical cyclone Pam in the Pacific left the island state of Vanuatu in ruins. While the death toll was thankfully much lower than in Haïti, with 16 killed, over half of the population was affected and much of the country's infrastructure destroyed. Estimated economic damage reached $450 million—equivalent to 64% of Vanuatu's gross domestic product (GDP) (Esler 2015). We could list more extreme weather events, such as floods and droughts, as well as slow-onset processes that are less likely to make headlines but are potentially just as devastating, such as ocean acidification, coral bleaching, or coastal erosion as a result of sea-level rise.[1]

The list of climate change impacts is long. While it is impossible to attribute any single event to global climate change, climate change does increase the frequency and intensity of weather extremes as well as contributes to changes like coastal erosion and flooding (IPCC 2014). Devastating disasters like the ones listed above are likely the 'new normal', a glimpse into 'a climate future that is less predictable and more extreme'

C. Betzold and F. Weiler, *Development Aid and Adaptation to Climate Change in Developing Countries*,
https://doi.org/10.1007/978-3-319-64510-0_1

(Byanyima et al. 2016). Coping with and adapting to this climate future is becoming more and more important and urgent around the globe—and nowhere more so than in the countries of the Global South. Developing countries not only feel the brunt of climate change impacts, as the weather extremes mentioned earlier illustrate, they also have fewer resources and less capacity to cope with and adapt to these adverse effects—and they have at the same time historically contributed the least to anthropogenic climate change (Mertz et al. 2009).

In light of this 'double injustice' of climate change, developed countries have a moral as well as a legal obligation to help vulnerable developing countries adapt. The 1992 United Nations Framework Convention on Climate Change (the Convention, or UNFCCC) requires developed countries to assist financially particularly vulnerable developing countries; more recent agreements, including the 2016 Paris Agreement, confirm this obligation and specifically call on developed countries to increase the level of financial assistance for adaptation (see Chap. 2 for a detailed overview).

This is what this book is about: financial support for adaptation to climate change from developed countries to developing countries. More specifically, we focus on bilateral official development assistance (ODA) that targets climate change adaptation in recipient countries—what we call adaptation aid. As we explain in greater detail below, bilateral adaptation aid is an important subset of adaptation finance and includes: support for projects such as building seawalls and (re-)planting mangroves; building or enhancing rainwater tanks; developing maps of safe areas in the case of hurricanes and other storms (see Noble et al. 2014, 844ff).

Support for adaptation raises many questions. Which adaptation actions are funded through international adaptation aid, and why? How effectively and sustainably do these adaptation actions reduce recipients' vulnerability and increase their resilience? How much finance is needed and how much is available for adaptation measures in developing countries? To what extent are adaptation (aid) and development (aid) different, and how? How can we measure and monitor adaptation finance flows? Where does this finance come from, and where does it go (e.g. Peterson Carvalho and Terpstra 2015)? In this book, we examine this last issue of distribution at the national level. Which countries receive how much support for adaptation, and why? How do donors allocate their adaptation aid across recipient countries? To what extent are the most vulnerable countries prioritised in adaptation aid allocation, as developed countries agreed and as we would expect from a climate justice perspective (cf. Barrett 2014)? We address these questions through an analysis of bilateral aid that targets adaptation to climate change, with a focus on three large climate donors: Germany, Sweden, and the UK.

In the remainder of this introductory chapter, we discuss two key concepts of this book, namely (i) adaptation to climate change and (ii) vulnerability and its components of exposure, sensitivity, and adaptive capacity. Both concepts are rather vague and lack a single definition. We then turn to the question of how much it will cost to adapt to climate change in developing countries, and where the funding necessary to pay for this adaptation will come from. We show that putting a price tag on adaptation is difficult. Despite considerable uncertainty as to the exact costs, developing countries will likely need tens of billions of dollars every year to deal with the impacts of climate change. This funding has to come from various sources, including donors. Developed countries are morally and legally obliged to support adaptation measures in developing countries, although the latter also contribute to covering the cost of adaptation. We will conclude this introductory chapter with an overview of the individual chapters of this book as well as a summary of its empirical, theoretical, and policy contributions.

1.1 Definition of Key Concepts

Before we move to defining the key concepts of adaptation and vulnerability, let us briefly clarify our aid terminology. In this book, we examine bilateral ODA, or aid. According to the Organisation for Economic Co-operation and Development (OECD), aid includes

> grants or loans to countries and territories on the DAC List of ODA Recipients (developing countries) and to multilateral agencies which are: (a) undertaken by the official sector; (b) with promotion of economic development and welfare as the main objective; (c) at concessional financial terms (OECD 2016).

We follow the OECD and define developing countries as those countries eligible to receive aid—that is, countries listed on the OECD Development Assistance Committee (DAC) List of ODA Recipients. This List includes countries based on gross national income (GNI) per capita and is revised every 3 years.[2] Countries on the list are potential recipients of adaptation aid; we consider all others to be developed or industrialised. We include as donors only those developed countries that report their ODA to the OECD.

1.1.1 *Adaptation to Climate Change*

Humans have always adapted to natural climate *variability*, that is, to the climatic conditions in which they live. Adaptation to anthropogenic climate *change* is different, at least in theory, notably because of different responsibilities: while adaptation to climate *variability* falls within the domestic responsibility of states, adaptation to climate *change* involves some level of globally shared responsibility (Burton 2004, 28). In practice, however, these two types of adaptation are hard to distinguish, particularly in developing countries, where the adaptation deficit is large: developing countries are already ill-prepared for current natural climate variability, let alone for the additional challenges posed by climate change (Burton 2004; Fankhauser 2010; Weikmans 2012).

How can we deal with climate change and climate variability? We have already mentioned some examples of concrete adaptation measures: building seawalls and (re-)planting mangroves, building or enhancing rainwater tanks, or developing maps of safe areas in the case of hurricanes and other storms are all measures to deal with different effects of climate change such as sea-level rise, changes in precipitation and the risk of drought, or more frequent and more intense weather extremes. More generally, the IPCC defines adaptation as the 'process of adjustment to actual or expected climate and its effects' (IPCC 2013, 1758). For the OECD, adaptation activities 'reduce the vulnerability of human or natural systems to the impacts of climate change and climate-related risks, by maintaining or increasing adaptive capacity and resilience' (OECD 2011, 4). While mitigation—reducing greenhouse gas emissions—is clearly defined, there are many ways to conceptualise adaptation (see Adger 2006; Hinkel 2011). This range of conceptualisations is somewhat problematic for the purpose of this book, as we discuss in Chap. 2.

1.1.2 *Vulnerability*

The purpose of adaptation is to reduce vulnerability. Just like adaptation, 'vulnerability' is complex and contested and has been conceptualised in different ways (Adger 2006; Füssel 2007; Muccione et al. 2017). Broadly

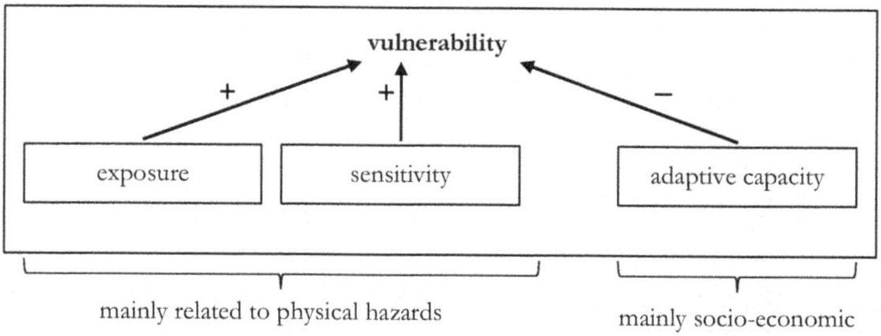

Fig. 1.1 Vulnerability and its components

speaking, vulnerability is the 'propensity or predisposition to be adversely affected' (IPCC 2013, 1775) or the 'susceptibility to be harmed' (Adger 2006, 269). The Fourth Assessment Report of the IPCC adds response capacity and defines vulnerability to climate change as 'the propensity of human and ecological systems to suffer harm and their ability to respond to stresses imposed as a result of climate change effects' (IPCC 2007, 720).

Vulnerability has three dimensions: it is a function of a system's exposure and sensitivity to perturbations or hazards, as well as the system's capacity or resilience to cope with, adapt to, and recover from the effects of these perturbations or hazards (see Fig. 1.1; Adger 2006; Smit and Wandel 2006). These elements of exposure, sensitivity, and adaptive capacity are: interrelated; vary over time, by type and according to stimulus; and are place-specific and system-specific (Smit and Wandel 2006, 286f).

Exposure and sensitivity are related concepts; Smit and Wandel (2006, 286) even see them as 'almost inseparable'. Exposure describes the 'nature and degree to which a system experiences environmental or socio-political stress', while sensitivity is 'the degree to which a system is modified or affected by perturbations' (Adger 2006, 270). More specifically for climate change, the IPCC defines exposure as the 'presence of people, livelihoods, species or ecosystems, environmental functions, services, and resources, infrastructure, or economic, social, or cultural assets in places and settings that *could be adversely affected*' and sensitivity as the 'degree to which a system or species *is affected*, either adversely or beneficially, by climate variability or change' (IPCC 2013, 1765; 1772; emphasis added). As Gallopín (2006, 296) highlights, sensitivity is an attribute of the system and independent of the perturbation, while exposure is an attribute of the relationship between the system and the perturbation.

Like sensitivity, adaptive capacity—the third element of vulnerability—is a system attribute that exists independent of the perturbation (Gallopín 2006, 296). Adger (2006, 270) defines adaptive capacity as the 'ability of a system to evolve in order to accommodate environmental hazards or policy change and to expand the range of variability with which it can cope' (see also Gallopín 2006; Smit and Wandel 2006). The IPCC definition is similar: 'The ability of systems, institutions, humans, and other organisms to adjust to potential damage, to take advantage of opportunities, or to respond to consequences' (IPCC 2013, 1758). Adaptive capacity is sometimes also referred to as adaptability, coping capacity, or resilience (Gallopín 2006). Others see these concepts as different but related: coping capacity refers to the shorter-term, adaptive capacity to longer-term responses; both are components of resilience (Gallopín 2006; Turner et al. 2003). Resilience is thus not the opposite, but a subset of vulnerability. Vulnerability—through its adaptive capacity component—not only depends on physical impact, but also on social, economic, institutional, and political structures and resources (Smit and Wandel 2006; Turner et al. 2003), as well as on access or entitlement to these structures and resources (Adger et al. 2003).

The three elements of vulnerability and their determinants are interrelated and interdependent (Füssel 2007); 'vulnerability to environmental change does not exist in isolation from the wider political economy of resource use' (Adger 2006, 270). A farmer in a semi-arid area whose fields are rainwater-fed, for example, may be *exposed* to climate change risks insofar as precipitation patterns change in the area where he lives and drought becomes more likely. The farmer is also *sensitive* to climate change risks because his livelihood—farming—depends on regular rainfall; changes in precipitation and more frequent droughts threaten his livelihood. The farmer's *adaptive capacity* depends on resources and his access to those resources. He may thus switch to more drought-resistant crops, improve irrigation, or diversify his livelihood through other income-generating activities. The fields of a wealthier farmer may be just as exposed, but he may be less vulnerable: if the farmer has a secure income from non-farming activities, cultivates diverse crops, and bought weather insurance, he is less sensitive to climate risks. At the same time, he has more adaptive capacity because he can more easily switch to other crops, install irrigation technology, or purchase insurance than a poorer farmer who has only rain-fed fields and no funds to purchase drought-resistant crops or insurance (see Mertz et al. 2009).

In some cases, there are limits to adaptation: it may simply not be feasible, or too expensive, to adapt to the effects of climate change. This is called residual damage, or loss and damage in the context of the climate change negotiations. Warner and van der Geest (2013, 369) define loss and damage as the 'negative effects of climate variability and climate change that people have not been able to cope with or adapt to'. This definition covers both the 'the inability to respond adequately to climate stressors' as well as economic or non-economic costs and adverse impacts of adaptation measures themselves (Warner and van der Geest 2013, 369).

Finally, a short notice on the use of terminology in this book. Because exposure and sensitivity are so closely linked with each other, we mostly discuss them jointly. Therefore, when we talk about the physical components of climate change vulnerability, we usually use the terms 'physical exposure and sensitivity' or simply 'physical vulnerability'.

1.2 The Cost of Climate Change Adaptation

It is difficult to estimate the costs of adaptation measures, including the costs of any adverse impacts of these measures themselves. Climate change adaptation is clearly 'not a costless exercise' (Adger et al. 2003, 191), but putting a price tag on it is extremely difficult. Several studies have tried to estimate the costs of adaptation since the 1990s, but because of different assumptions, different methods, and different coverages, the estimates vary widely, ranging from $4 billion to over $100 billion per year (Weikmans 2012, para. 2; Narain et al. 2011).

The first set of studies that sought to estimate the costs of climate change adaptation focused on calculating not directly the cost of adaptation, but the cost of climate change impacts (e.g. Nordhaus 1995; Tol 1995; see Weikmans 2012, para. 13 and Fankhauser 2010 for reviews). Given the growing interest in adaptation and climate finance (see Chap. 2), later studies sought to estimate the cost of adaptation specifically for developing countries. These studies used two approaches. One approach identifies the fraction of current investments that is climate sensitive, and then applies a 'mark up' factor that reflects the cost of 'climate-proofing' those investments (Smith et al. 2011, 988; Fankhauser 2010). The other approach computes the cost of the adaptation projects listed in individual

National Adaptation Programmes of Action (NAPAs) and scales these costs up to all developing countries (Smith et al. 2011, 988). The costs of adaptation, according to these approaches, differ widely (Smith et al. 2011, 988). Annual adaptation costs lie between $9 billion and $41 billion according to a report by the World Bank (2006) and between $4 billion and $37 billion in the Stern Review (Stern 2007, 501f; the report directly builds on the World Bank report). A study by Oxfam puts the price tag at over $50 billion per year (Oxfam International 2007), while the United Nations Development Programme (UNDP) estimates that at least $44 billion are needed for climate-proofing development investments and infrastructure, plus $42 billion for adapting poverty reduction and strengthening disaster response (UNDP 2007, 194).

The second generation of estimates refined the methods used and sought to identify the costs of adaptation in different sectors (Fankhauser 2010). The studies by the UNFCCC and Project Catalyst come to fairly similar costs of between $28 billion and $67 billion per year in 2030 (UNFCCC 2007), and between $26 billion and $77 billion (Project Catalyst 2009), respectively. According to the latest report by the World Bank, between $70 billion and $100 billion will be needed every year between 2010 and 2050 to adapt to a world that is 2 °C warmer in 2050 (World Bank 2010; Narain et al. 2011).

As the wide range of estimates shows, there is a lot of uncertainty involved in putting a price tag on adaptation. Reviewers as well as authors of studies that estimate adaptation costs emphasise gaps and limitations, including the scope and depth of the analysis (for instance, which climate impacts and which adaptation measures are considered), the costing of measures, or the treatment of uncertainty (Fankhauser 2009, 25f, 2010; Agrawala and Fankhauser 2008; Weikmans 2012). Few studies, for instance, factor in 'soft' adaptation measures such as capacity building or planning since it is more difficult to assess the costs of these measures (Narain et al. 2011, 1004), and no study considers the cost of loss and damage—situations where adaptation is not feasible or too expensive (Smith et al. 2011, 990; see above). Additionally, estimates focus on the additional cost of adaptation to man-made climate change only, and disregard the cost of adaptation to natural climate change variability, even if developing countries in particular are generally ill-equipped to deal with even current climate variability—the adaptation gap or deficit mentioned earlier (Burton 2004; Weikmans 2012). Finally, it is also unclear what exactly counts as adaptation, since it is difficult, if not impossible, to disentangle it from development more broadly, as we discuss in greater detail in Chap. 2. Let it suffice to say here that, overall, tens of billions of dollars will be needed every year to deal with the effects of climate change and that true adaptation costs are likely to be higher than the estimated costs (Fankhauser 2009, 26).

1.3 Financing Climate Change Adaptation

In light of this uncertainty, how should and how do developing countries raise these tens of billions of dollars every year? And to what extent do available funds meet adaptation needs? To cover costs, different sources of funding will be important, both public and private, national and international (see Bouwer and Aerts 2006). First, developing countries cannot—and should not—pay for all adaptation expenses themselves. Adaptation to climate change has an international dimension and all countries share a global responsibility to fund it (Burton 2004). From a climate justice perspective and following the polluter pays principle, developed countries, whose emissions have historically contributed the most to anthropogenic climate change, have an obligation to (co-)fund adaptation in developing countries, which will and already do feel the brunt of climate change impacts. Developed countries have accepted this responsibility by agreeing to assist 'particularly vulnerable' developing countries to meet the cost of adaptation in the UNFCCC (UNFCCC 1992, Article 4.4; see Chap. 2). This means that developed countries have a legal obligation to help at least some developing countries adapt to climate change. Yet, this obligation is rather vague—the agreement, for instance, does not specify what this assistance should look like or how much assistance is required. More recently, developed countries put forward concrete numbers and agreed to provide \$30 billion in so-called 'fast-start finance' between 2010 and 2012, and pledged to 'mobilise' \$100 billion per year by 2020 (UNFCCC 2009, para. 8; UNFCCC 2010, Decision 1/CP.16, para. 98). We discuss the issue of adaptation and adaptation finance in global climate change negotiations in greater detail in Chap. 2 (Sect. 2.1); here, we focus on the different sources of finance, including those raised as a result of climate change negotiations.

Both figures—the \$30 billion 'fast-start finance' and the \$100 billion target—include funding for adaptation and mitigation. In practice, most of the finance so far has been invested in mitigation and only a small share went to adaptation. The *Landscape of Climate Finance* reports have tracked public and private investment in low carbon and climate resilient actions globally since 2011 (Buchner et al. 2011, 2012, 2013, 2014, 2015). According to their estimates, only a small but growing share of financial flows went into adaptation activities. In 2011, only 4.5% of the flows covered targeted adaptation (Buchner et al. 2011); in 2015, this figure went up to 17% (Buchner et al. 2015). The UNFCCC Standing Committee on Finance (2014, 2016) has produced two biennial assessment reports of climate finance flows. These assessments estimate that about one-quarter of all flows has been spent on adaptation activities across the 4 years assessed (2011 through 2014). A joint report by the Climate Policy Initiative—the authors of the *Landscape of Climate Finance* reports—and the OECD sought to measure more precisely climate finance flows that count towards the $100 billion target (OECD 2015). The report puts total climate finance flows at $52 billion in 2013 and at $62 billion in 2014, of which 16% targeted adaptation and 7% adaptation and mitigation. These last estimates, however, were strongly criticised. The Indian government reassessed the OECD numbers and concluded that only $2.2 billion in new and additional resources had been disbursed, though the report does not specify how much of this went into adaptation (Climate Change Finance Unit 2015). Even if different studies come to different conclusions with regard to the precise scale of climate and adaptation finance, they agree insofar as 'none finds that a sufficient amount is currently being mobilized to meet the climate challenge' (Ha et al. 2016, 2).

If we leave aside the question of what is new and additional and what exactly counts as adaptation as opposed to development (see Chap. 2), relatively reliable and comparable numbers are easier to find, in particular for public funding. Public sources of funding are more relevant for adaptation than for mitigation, since there are fewer gains to be made from investments in adaptation compared to for instance renewable energy (Atteridge 2011). Developed countries provide almost all of their funding for adaptation from aid budgets (e.g. Ayers and Abeysinghe 2013; Weikmans 2016); only a small part of adaptation finance is specifically raised for this purpose. The most prominent example here is the Adaptation Fund, a multilateral fund established under the UNFCCC (see Chap. 2). The Adaptation Fund is partly resourced through a levy on the so-called Clean Development Mechanism, which allows developed countries to reduce their greenhouse gas emissions by investing in mitigation activities in developing countries.[3]

Donors can provide aid either through bilateral channels or through multilateral funds such as the Adaptation Fund. In practice, donors prefer bilateral over multilateral channels, not least because they have greater control over where and on what their aid is spent (Gulrajani 2016). On the other hand, some donors, notably small ones, provide a majority of their funding multilaterally (Gulrajani 2016; see also Chap. 4). Although a number of multilateral funds have specifically been established to assist developing countries with climate change adaptation (and mitigation), the majority of adaptation aid flows through bilateral channels. For the $30 billion fast-start finance provided between 2010 and 2012, Weikmans (2016) reports that 60% of all climate finance was given bilaterally, 30% through traditional multilateral actors like the World Bank, and only 10% through multilateral funds under the UNFCCC.

The focus of this book is on bilateral adaptation aid as the largest share of adaptation finance to date. Nonetheless, we would like to conclude this section with a reference to South–South and domestic adaptation finance. Even if North–South adaptation finance flows are important from a justice perspective, we should not forget that developing countries themselves are spending increasing amounts of funding on adaptation, either at home or—to a lesser extent—in other developing countries. It is hard to get numbers on this funding (Ha et al. 2016). From the limited data available for a few developing countries, the 2016 Biennial Assessment of the UNFCCC Standing Committee on Finance (2016, 6) concludes that 'in these countries, domestic public finance significantly exceeds the inflows of international public climate finance from bilateral and multilateral sources'. Additionally, some developing countries[4] have started to provide climate finance to other developing countries, bilaterally, as contributions to established multilateral funds, or as contributions to new Southern organisations like the 'BRICS bank' (Ha et al. 2016). Again, however, data is limited. Estimates put South–South climate finance at around $6 billion to $12 billion, but it is unclear how much of this targets adaptation as compared to mitigation (see UNFCCC Standing Committee on Finance 2016, 52).

1.4 Overview of the Book and Contributions

In this book, we analyse a subset—if an important one—of global adaptation finance flows, namely bilateral adaptation aid, and especially focus on adaptation aid from three large climate donors: Germany, Sweden, and the UK. We seek to understand how donors distribute their adaptation aid across recipient countries.

We have already laid out the broad background of our analysis: the growing need for adaptation in developing countries to reduce vulnerability to climate change and variability, and the obligation and commitment of developed countries to assist developing countries in their efforts to deal with the effects of climate change. In the next chapter, we will deepen this background. We first show how adaptation has risen on the agenda of the UNFCCC negotiations. While adaptation was for a long time the 'little brother' of mitigation, adaptation and adaptation finance have become more and more important over the course of the climate negotiations. The 2015 Paris Agreement includes a global adaptation goal and recognises adaptation as equally important as mitigation.

The second half of Chap. 2 then reviews the literature on aid allocation. Scholars distinguish between three major determinants of aid allocation: recipient need, recipient merit, and donor interests. According to the recipient need model, donors altruistically give their aid to the poorest countries that most need support to develop. According to the recipient merit model, donors reward countries with good governance and the 'right' policies, where aid also tends to be more effective. According to the donor interest models, donors use aid to promote their own economic and political interests. These determinants likely also play a role when donors distribute their aid for climate change adaptation. A number of studies have empirically examined the distribution of adaptation aid and mostly confirmed the empirical results of the broader development aid literature: donors have different motivations when allocating their aid, with donor interests often dominating.

We empirically assess the three determinants of aid allocation in the context of climate change adaptation. Chapter 3 lays out our research design, data, and methods of analysis. We opted for a mixed methods research design that combines quantitative analyses of OECD aid data with qualitative case studies of Germany, Sweden, and the UK based on semi-structured interviews and policy documents. As Chap. 3 explains, this research design allows us to trace adaptation aid from 2010 through 2015 across the developing world, as well as to look 'behind the numbers' and to understand better how aid allocation decisions are made and what role climate change and adaptation play in development cooperation in these three countries. The qualitative analysis also helps overcome inherent weaknesses of aid data and statistical analyses, as we explain (see also Roodman 2007).

Chapters 4 through 6 present the results of our empirical analysis. Chapter 4 starts with a descriptive analysis of the OECD data. The OECD introduced a so-called adaptation marker in 2009. Since 2010, donors have been required to indicate which parts of their development aid are relevant for climate change adaptation. While self-reporting is always problematic and the OECD data are far from perfect, they represent the most comprehensive and comparable data that is available to date. We therefore use the adaptation marker to explore adaptation aid over time and across space. We examine, among other things, how much aid targets adaptation, how this differs across donors, and how adaptation aid is distributed globally. The analysis shows that adaptation aid is still a rather small share of total development aid, but is increasing in importance. Most adaptation is disbursed rather quickly and mainly distributed through bilateral channels. In absolute terms, Japan and Germany are the largest contributors to adaptation aid, while the Scandinavian countries including Sweden are the most generous providers of it in per capita terms. At first glance, vulnerability matters for how donors allocate their adaptation aid: populous and rather vulnerable South East Asian countries like Vietnam, the Philippines, and Bangladesh are among the top recipients of adaptation aid in absolute terms, while some very small and highly vulnerable SIDS like Tuvalu and Niue are among the top recipients in per capita terms.

Chapter 5 turns to a more systematic assessment of the geographic distribution of adaptation aid. We model the allocation decision as a two-stage process: first, donors need to decide to which countries they want to provide adaptation aid, and, in a second step, how much of it they want to provide to each selected recipient. We examine the role of the three determinants—recipient need or vulnerability to climate change, recipient merit, and donors' own political and economic interests—at each of these two stages. We first do so for all donors together. In a second step, we disaggregate our analysis further and examine adaptation aid allocation separately for our three case study countries: Germany, Sweden, and the UK. The results of the statistical analyses indicate that recipient need, recipient merit, and donor interests all guide donors when they allocate adaptation aid. Countries that are physically more vulnerable to climate change, well-governed countries, and countries to which donors export a lot tend to receive more adaptation aid, all else being equal. Notably, however, we also find that adaptation aid is very closely linked to development aid in general: donors support adaptation in those countries in which they have been engaged in development cooperation previously. This path dependency puts into question the additionality of adaptation aid. We find similar factors at play for our case study countries, though the results are generally weaker and point to different emphases. Sweden, for instance, pays particular attention to good governance, while the United Kingdom considers physical exposure to climate change.

Chapter 6 delves even deeper into the adaptation aid allocation processes in these three countries. Based on our interviews with aid practitioners and observers as well as an analysis of key policy documents, we review the overall aid architecture and decision-making processes in these countries and the role of climate change and adaptation in development cooperation more broadly. We then turn to the three determinants of aid allocation and investigate how these determinants come into play in the context of adaptation aid. All three countries consider climate change and increasingly adaptation as priority areas for development cooperation. They stress their commitment to supporting vulnerable countries in particular, but also note that it is difficult to identify vulnerable countries, to distinguish adaptation from development more generally, and to engage in development cooperation and adaptation projects in the poorest and least developed countries where capacity is lacking. Finally, we address additional factors and issues that are central in policy debates but that we cannot detect in our quantitative analysis, such as the distinction between development assistance and adaptation finance, the role of public opinion, or issues of reporting adaptation aid.

Finally, the concluding chapter (Chap. 7) draws together our findings from the empirical analyses. We summarise again our research design as well as the empirical results, and discuss the empirical, theoretical, and policy implications of our study. Empirically, our study provides the most comprehensive analysis of adaptation aid allocation to date, based on both a dyadic dataset covering all donors and all recipients included in the OECD Creditor Reporting System and qualitative in-depth analysis of three large climate donors. Theoretically, our study complements the broader aid allocation literature. We show that recipient need is a concept that is broader than just poverty and specific to the type of aid studied. In the context of climate change adaptation, recipient need translates into vulnerability to climate risks. Finally, in terms of policy, our results imply both good news and bad news. The good news is that vulnerability does matter: donors do support adaptation in particularly vulnerable countries, as they repeatedly promised in the climate change negotiations. However, donors focus on physical exposure and sensitivity rather than adaptive capacity, not least because aid is presumably less effective in countries with low adaptive capacity. The bad news is that adaptation aid by and large follows development aid. In other words, if a donor has provided some form of development aid to a recipient, the donor is very likely also to provide adaptation aid to this recipient—regardless of the recipient's level of vulnerability or its governance. This link is unsurprising insofar as adaptation aid is of course a subset of development aid. But adaptation aid also contributes to adaptation finance, which must be new and additional to 'regular' development aid. If donors are to keep their promise of providing ***additional*** resources for the ***additional*** burden that climate change poses to developing countries, they must provide more than just adaptation aid, and they must support all vulnerable developing countries, not just their traditional development partners.

Notes

1. The Fifth Assessment Report of the Intergovernmental Panel on Climate Change (IPCC) provides the most comprehensive overview of climate change impacts to date (see the summary for policymakers, IPCC 2014).
2. See http://www.oecd.org/dac/stats/historyofdaclistsofaidrecipientcountries.htm for more details.
3. For more information on the Adaptation Fund, see the Adaptation Fund's website at http://adaptation-fund.org/.
4. This includes countries that are relatively rich but do not have financial commitments under the UNFCCC such as South Korea or the United Arab Emirates (which report their aid to the OECD), but also for instance China (see UNFCCC Standing Committee on Finance 2016, 52).

References

Adger, W.N. 2006. Vulnerability. *Global Environmental Change* 16 (3): 268–281.

Adger, W.N., S. Huq, K. Brown, D. Conway, and M. Hulme. 2003. Adaptation to Climate Change in the Developing World. *Progress in Development Studies* 3 (3): 179–195.

Agrawala, S., and S. Fankhauser, eds. 2008. *Economic Aspects of Adaptation to Climate Change: Costs, Benefits and Policy Instruments*. Paris: OECD Publishing.

AIR Worldwide. 2016. AIR Worldwide Estimates Combined Insured Losses for Hurricane Matthew for the United States and the Caribbean. Press Release, 13 October 2016. Available online at http://www.air-worldwide.com/

Press-Releases/AIR-Worldwide-Estimates-Combined-Insured-Losses-for-Hurricane-Matthew-for-the-United-States-and-the-Caribbean/.

Atteridge, A. 2011. Will Private Finance Support Climate Change Adaptation in Developing Countries? Historical Investment Patterns as a Window on Future Private Climate Finance. Stockholm Environment Institute, Working Paper 2011.

Ayers, J.M., and A.C. Abeysinghe. 2013. International Aid and Adaptation to Climate Change. In *The Handbook of Global Climate and Environmental Policy*, chap. 28, ed. R. Falkner, 486–506. New York: Wiley.

Barrett, S. 2014. Subnational Climate Justice? Adaptation Finance Distribution and Climate Vulnerability. *World Development* 58: 130–142.

Bouwer, L.M., and J.C. Aerts. 2006. Financing Climate Change Adaptation. *Disasters* 30 (1): 49–63.

Buchner, B., A. Falconer, M. Hervé-Mignucci, C. Trabacchi, and M. Brinkman. 2011. The Landscape of Climate Finance: A CPI Report. Climate Policy Initiative, Venice.

Buchner, B., A. Falconer, M. Hervé-Mignucci, and C. Trabacchi. 2012. The Landscape of Climate Finance 2012. Climate Policy Initiative, Venice.

Buchner, B., M. Hervé-Mignucci, C. Trabacchi, J. Wilkinson, M. Stadelmann, R. Boyd, F. Mazza, A. Falconer, and V. Micale. 2013. The Global Landscape of Climate Finance 2013. Climate Policy Initiative, Venice.

Buchner, B., M. Stadelmann, J. Wilkinson, F. Mazza, A. Rosenberg, and D. Abramskiehn. 2014. The Global Landscape of Climate Finance 2014. Climate Policy Initiative, Venice.

Buchner, B., C. Trabacchi, F. Mazza, D. Abramskiehn, and D. Wang. 2015. Global Landscape of Climate Finance 2015. Climate Policy Initiative, Venice.

Burton, I. 2004. Climate Change and the Adaptation Deficit. In *Climate Change: Building the Adaptive Capacity: An International Conference on Adaptation Science, Management and Policy Options*, ed. A. Fenech, D. MacIver, H. Auld, R.B. Rong, and Y. Yin, 25–33. Toronto: Meteorological Service of Canada, Environment Canada.

Byanyima, W., M. Robinson, and M. Kamau. 2016. This El Niño Is A Huge Humanitarian Crisis - But It Didn't Have To Be. The Huffington Post, 23 September 2016. http://www.huffingtonpost.com/winnie-byanyima/this-el-nino-is-a-huge-hu_b_12137892.html. Accessed 4 Nov 2016.

Dasgupta, D., and Climate Change Finance Unit. 2015. Climate Change Finance, Analysis of a Recent OECD Report: Some Credible Facts Needed. Climate Change Finance Unit, Department of Economic Affairs, Ministry of Finance, Government of India. Available online at http://pibphoto.nic.in/documents/rlink/2015/nov/p201511290l.pdf.

Esler, S. 2015. Vanuatu Post-Disaster Needs Assessment. Tropical Cyclone Pam, March 2015. Government of Vanuatu, Port Vila. Available online at https://www.gfdrr.org/sites/default/files/publication/PDNA_Cyclone_Pam_Vanuatu_Report.pdf.

Fankhauser, S. 2009. The Range of Global Estimates. In *Assessing the Costs of Adaptation to Climate Change: A Review of the UNFCCC and Other Recent Estimates*, chap. 1, ed. M. Parry, N. Arnell, P. Berry, D. Dodman, S. Fankhauser, C. Hope, S. Kovats, R. Nicholls, D. Satterthwaite, R. Tiffin, and T. Wheeler, 19–28. London: International Institute for Environment and Development and Grantham Institute for Climate Change.

Fankhauser, S. 2010. The Costs of Adaptation. *WIRES Climate Change* 1: 23–30.

Füssel, H.-M. 2007. Vulnerability: A Generally Applicable Conceptual Framework for Climate Change Research. *Global Environmental Change* 17: 155–167.

Gallopín, G.C. 2006. Linkages between Vulnerability, Resilience, an Adaptive Capacity. *Global Environmental Change* 16: 293–303.

Gulrajani, N. 2016. Bilateral Versus Multilateral Aid Channels: Strategic Choices for Donors. ODI Report. Available online at https://www.odi.org/sites/odi.org.uk/files/resource-documents/10492.pdf, Overseas Development Institute, London.

Ha, S., T. Hale, and P. Ogden. 2016. Climate Finance in and between Developing Countries: An Emerging Opportunity to Build On. *Global Policy* 7 (1): 102–108.

Hinkel, J. 2011. "Indicators of vulnerability and adaptive capacity": Towards a Clarification of the Science–Policy Interface. *Global Environmental Change* 21: 198–208.

IPCC. 2007. Assessment of Adaptation Practices, Options, Constraints and Capacity. In *Climate Change 2007: Impacts, Adaptation and Vulnerability. Contribution of Working Group II to the Fourth Assessment Report of the Intergovernmental Panel on Climate Change*, chap. 17, ed. M. Parry, O. Canziani, J. Palutikof, P. van der Linden, and C. Hanson, 717–743. Cambridge: Cambridge University Press.

IPCC. 2013. Annex III: Glossary. In *Climate Change 2013: The Physical Science Basis. Contribution of Working Group I to the Fifth Assessment Report of the Intergovernmental Panel on Climate Change*, ed. T. Stocker, D. Qin, G.-K. Plattner, M. Tignor, S. Allen, J. Boschung, A. Nauels, Y. Xia, V. Bex, and P. Midgley, 1447–1466. Cambridge/New York: Cambridge University Press.

IPCC. 2014. Summary for Policymakers. In *Climate Change 2014: Impacts, Adaptation, and Vulnerability. Part A: Global and Sectoral Aspects. Contribution of Working Group II to the Fifth Assessment Report of the Intergovernmental Panel on Climate Change*, ed. C. Field, V. Barros, D. Dokken, K. Mach, M. Mastrandrea, T. Bilir, M. Chatterjee, K. Ebi, Y. Estrada, R. Genova, B. Girma, E. Kissel, A. Levy, S. MacCracken, P. Mastrandrea, and L. White, 1–32. Cambridge/New York: Cambridge University Press.

Mertz, O., K. Halsnæs, J.E. Olesen, and K. Rasmussen. 2009. Adaptation to Climate Change in Developing Countries. *Environmental Management* 43 (5): 743–752.

Muccione, V., S.K. Allend, C. Huggel, and J. Birkmann. 2017. Differentiating Regions for Adaptation Financing: The Role of Global Vulnerability and Risk Distributions. *WIREs Climate Change* 8 (2): e447.

Narain, U., S. Margulis, and T. Essam. 2011. Estimating Costs of Adaptation to Climate Change. *Climate Policy* 11 (3): 1001–1019.

Noble, I., S. Huq, Y. Anokhin, J. Carmin, D. Goudou, F. Lansigan, B. Osman-Elasha, and A. Villamizar. 2014. Adaptation Needs and Options. In *Climate Change 2014: Impacts, Adaptation, and Vulnerability. Part A: Global and Sectoral Aspects. Contribution of Working Group II to the Fifth Assessment Report of the Intergovernmental Panel on Climate Change*, chap. 14, ed. C. Field, V. Barros, D. Dokken, K. Mach, M. Mastrandrea, T. Bilir, M. Chatterjee, K. Ebi, Y. Estrada, R. Genova, B. Girma, E. Kissel, A. Levy, S. MacCracken, P. Mastrandrea, and L. White, 833–868. Cambridge/New York: Cambridge University Press.

Nordhaus, W. 1995. *Managing the Global Commons: The Economics of Climate Change*. Cambridge, MA: The MIT Press.

OCHA. 2016. Haiti: Hurricane Matthew. Situation Report No. 19 (2 November 2016). http://reliefweb.int/sites/reliefweb.int/files/resources/ocha_haiti_sitrep_19_02_nov_2016_eng.pdf. Accessed 4 Nov 2016.

OECD. 2011. Handbook on the OECD-DAC Climate Markers. Available online at www.oecd.org/dac/stats/48785310.pdf.

OECD. 2015. Climate Finance in 2013–14 and the USD 100 Billion Goal. A Report by the OECD in collaboration with Climate Policy Initiative. Available online at http://www.oecd.org/environment/cc/OECD-CPI-Climate-Finance-Report.htm.

OECD. 2016. DAC Glossary of Key Terms and Concepts. Available online at http://www.oecd.org/dac/dac-glossary.htm. Accessed 9 Nov 2016.

Oxfam International. 2007. Adapting to Climate Change: What's Needed in Poor Countries, and Who Should Pay. Oxfam Briefing Paper 104. Available online at https://www.oxfam.org/sites/www.oxfam.org/files/adapting%20to%20climate%20change.pdf.

Peterson Carvalho, A., and P. Terpstra. 2015. Tracking Adaptation Finance: An Approach for Civil Society Organizations to Improve Accountability for Climate Change Adaptation. Oxfam America Inc. and World Resources Institute. Available online at http://www.wri.org/publication/tracking-adaptation-finance.

Project Catalyst. 2009. Adaptation to Climate Change: Potential Costs and Choices for a Global Agreement. Findings of the Adaptation Working Group of Project Catalyst. San Francisco: Climate Works Foundation.

Roodman, D. 2007. The Anarchy of Numbers: Aid, Development, and Cross-country Empirics. Center for Global Development Working Paper Number 32.

Smit, B., and J. Wandel. 2006. Adaptation, Adaptive Capacity and Vulnerability. *Global Environmental Change* 16 (3): 282–292.

Smith, J.B., T. Dickinson, J.D.B. Donahue, I. Burton, E. Haites, R.J. Klein, and A. Patwardhan. 2011. Development and Climate Change Adaptation Funding: Coordination and Integration. *Climate Policy* 11: 987–1000.

Stern, N.H. 2007. *The Economics of Climate Change: The Stern Review*. Cambridge: Cambridge University Press.

Tol, R.S.J. 1995. The Damage Costs of Climate Change: Towards More Comprehensive Calculations. *Environmental and Resource Economics* 5: 353–374.

Turner, B.L., R.E. Kasperson, P.A. Matsone, J.J. McCarthy, R.W. Corell, L. Christensen, N. Eckley, J.X. Kasperson, A. Luers, M.L. Martello, C. Polsky, A. Pulsiphera, and A. Schiller. 2003. A Framework for Vulnerability Analysis in Sustainability Science. *Proceedings of the National Academy of Sciences* 100 (14): 8074–8079.

UNDP. 2007. Human Development Report 2007/2008. Fighting Climate Change: Human Solidarity in a Divided World. United Nations Development Programme, New York. Available online at http://hdr.undp.org/sites/default/files/reports/268/hdr_20072008_en_complete.pdf.

UNFCCC. 1992. United Nations Framework Convention on Climate Change. Document FCCC/INFORMAL/84.

UNFCCC. 2007. Investments and Financial Flows to Address Climate Change. Technical Report, United Nations Framework Convention on Climate Change Background Paper. Available online at http://unfccc.int/files/cooperation_and_support/financial_mechanism/application/pdf/background_paper.pdf, Bonn.

UNFCCC. 2009. Report of the Conference of the Parties on Its Fifteenth Session, Held in Copenhagen from 7 to 19 December 2009. Addendum. Part Two: Action Taken by the Conference of the Parties at its Fifteenth Session. Document FCCC/CP/2009/11/Add.1.

UNFCCC. 2010. The Cancún Agreements: Outcome of the Work of the Ad Hoc Working Group on Long-term Cooperative Action Under the Convention. Decision 1/CP.16. Document FCCC/CP/2010/7/Add.1.

UNFCCC Standing Committee on Finance. 2014. 2014 Biennial Assessment and Overview of Climate Finance Flows Report. Available online at http://unfccc.int/files/cooperation_and_support/financial_mechanism/standing_committee/application/pdf/2014_biennial_assessment_and_overview_of_climate_finance_flows_report_web.pdf. Accessed 29 Nov 2016.

UNFCCC Standing Committee on Finance. 2016. 2016 Biennial Assessment and Overview of Climate Finance Flows Report. Available online at http://unfccc.int/cooperation_and_support/financial_mechanism/standing_committee/items/6877.php.

Warner, K., and K. van der Geest. 2013. Loss and Damage from Climate Change: Local-Level Evidence from Nine Vulnerable Countries. *International Journal of Global Warming* 5 (4): 367–386.

Weikmans, R. 2012. Le coût de l'adaptation aux changements climatiques dans les pays en développement. *VertigO - la revue électronique en sciences de l'environnement* 12 (1).

Weikmans, R. 2016. Le rôle de la coopération au développement dans le financement international de l'adaptation au changement climatique. In *La nouvelle géographie du développement. Coopérer dans un monde en mutation*, ed. A. Zacharie, 175–185. Brussels: Éditions Le Bord de l'Eau.

World Bank. 2006. *Investment Framework for Clean Energy and Development.* Washington, DC: The World Bank.

World Bank. 2010. Economics of Adaptation to Climate Change: Synthesis Report. Technical Report, The International Bank for Reconstruction and Development/The World Bank, Washington, DC.

CHAPTER 2

The History and Political Economy of Adaptation Aid

In order to understand better how adaptation aid is disbursed, we need to know where this new form of aid comes from, how it is related to development aid, and what the theoretical expectations for adaptation aid allocation are. This chapter addresses these questions and thus sets the stage for the empirical analyses later on.

Adaptation aid—the focus of this book—is a subset of adaptation finance, which in turn is a subset of climate finance. Climate finance includes all types of 'financial flows from developed to developing countries for climate action' (Peterson Carvalho and Terpstra 2015, 6). Yet what exactly counts as climate finance is contested, as are many other questions. In this chapter, we first outline the history of adaptation, and specifically adaptation finance, in the international climate change negotiations (Sect. 2.1). We show how adaptation only appeared on the margins in the early period of climate change negotiations, but slowly became more and more important. Today, adaptation is at the centre of climate change negotiations, next to mitigation. Accordingly, adaptation finance has also become a central issue in the negotiations. Yet, how it is related to or different from development aid is a controversial question (Sect. 2.2). We do not seek to solve this question but start from the empirical observation that most adaptation finance to date has been given as development aid. Because development aid has been around for much longer than adaptation aid, there is a much larger literature on aid allocation more broadly. We review this literature in Sect. 2.3. Studies have identified three models

C. Betzold and F. Weiler, *Development Aid and Adaptation to Climate Change in Developing Countries*,
https://doi.org/10.1007/978-3-319-64510-0_2

of aid allocation: recipient need, recipient merit, and donor interests. Empirically, these three models coexist: donors use aid to promote poverty reduction (recipient need), to reward 'good' policies and institutions (recipient merit), and to further their own economic and political interest (donor interests)—although donor interests often seem to dominate. A smaller body of literature has tested to what extent these models apply to the allocation of adaptation aid (Sect. 2.4). This research finds that similar factors are at play, but that recipient need—which translates into vulnerability to climate change impacts in an adaptation context—does not play a large role. Possibly, this finding is also connected with the difficulties of measuring vulnerability, which is fundamentally about risk and, as such, subjective. We conclude this chapter with a brief summary and our expectations that build on the general aid allocation literature.

2.1 History of Adaptation and Adaptation Finance in Climate Change Negotiations

What is the history of adaptation aid? Where does this new form of aid come from? We need first to understand the historical developments that led to the establishment of climate and adaptation aid before we can explain how these funds are distributed. Financial support from developed countries to developing countries is a central element of many international environmental agreements, and climate change is no exception: 'At virtually all milestones in climate talks, promises of funding have been critical in breaking impasses: in Rio de Janeiro in 1992, in Kyoto in 1997, Marrakesh in 2001, Copenhagen in 2009 and Paris in 2015' (AdaptationWatch Weikmans et al., 30). Financial payments are an important instrument to bridge the North–South divide and address, to some extent, issues of equity and fairness. The next section traces the ascent of adaptation and specifically adaptation finance in the global climate negotiations. While adaptation was for a long time the 'little brother' of mitigation, its importance has steadily increased, not least because we do not mitigate enough. Adaptation is currently as important as mitigation.

2.1.1 Adaptation at the Sidelines: The Early Years of Climate Change Negotiations

The United Nations Framework Convention on Climate Change (UNFCCC) was signed at the United Nations Conference on Environment and Development, or Earth Summit, in Rio de Janeiro in 1992. The Convention rests on the principle of 'common but differentiated responsibilities and respective capabilities' (UNFCCC 1992, Preamble). As with other environmental agreements, the Convention makes a distinction between developed and developing countries and acknowledges that the former have historically contributed much more to anthropogenic climate change than the latter and at the same time have more resources to address climate change (see Harris 1999; Stone 2004). To take into account the different levels of greenhouse gas emissions and to address questions of equity and fairness, the Convention specifies different commitments for developed country Parties and developing country Parties. Developed countries as well as economies in transition are listed in Annex I to the Convention; these Annex I countries have specific reporting obligations about their greenhouse gas emissions (and binding emissions reduction obligations under the 1997 Kyoto Protocol). Developed countries—the then member of the Organisation of Economic Co-operation and Development (OECD) plus the then European Economic Community—are listed in Annex II. Financial commitments apply to these Annex II countries: they shall 'provide new and additional financial resources to meet the agreed full costs incurred by developing country Parties' for reporting as well as other commitments (UNFCCC 1992, Article 4.3). These other commitments include 'national, and where appropriate, regional programmes containing [...] measures to facilitate adequate adaptation to climate change' (UNFCCC 1992, Article 4.1). Article 4.4 further stipulates that the Annex II countries 'shall also assist the developing country Parties that are particularly vulnerable to the adverse effects of climate change in meeting costs of adaptation to those adverse effects' (UNFCCC 1992, Article 4.4). The Global Environment Facility (GEF), established in 1991, was designated as the interim financial mechanism of the Convention and later of the Kyoto Protocol (UNFCCC, 1992, Article 21.3, 1997, Article 11.2).

The first Conference of the Parties to the Convention (COP1) decided on adaptation measures as a result of pressure from developing countries. In the short term, this referred to 'planning [...] to identify particularly vulnerable countries or regions and policy options for adaptation and appropriate capacity-building', and in the medium to long term, to measures to prepare and facilitate adaptation (UNFCCC 1995, Decision 11/CP.1, para. 1(d)). Note the soft formulations: the decision talks about planning, preparing, and facilitating adaptation, but not (yet) about implementing concrete adaptation actions.

The 1997 Kyoto Protocol specified binding emission reduction targets for developed countries, yet 'adaptation received only lip service in two articles' (Gupta 2014, 91). Article 10.b confirms the Convention commitments to adaptation (UNFCCC 1997, Article 10.b). Additionally, a share of the proceeds of the Clean Development Mechanism is to be used to assist particularly vulnerable countries with adaptation (UNFCCC 1997, Article 12.8).

Yet, neither the Convention nor the Protocol defines who 'particularly vulnerable' developing countries are or how they should be determined. While the Preamble to the Convention recognises 'low-lying and other small island countries, countries with low-lying coastal, arid and semi-arid areas or areas liable to floods, drought and desertification, and developing countries with fragile mountainous ecosystems' as particularly vulnerable (UNFCCC 1992, Preamble), more or less all developing countries fit this description in one way or another (Harmeling and Kaloga 2011; Klein and Möhner 2011).

Further, the language on adaptation in the Convention, the first COP decisions, and the Kyoto Protocol is rather weak. The agreements talk about preparation, planning, and facilitating measures rather than concrete actions (see e.g. UNFCCC 1995, Decision 11/CP.1, para. 1(d); Burton et al. 2002; Khan and Roberts 2013). This language reflects the lower status of adaptation as compared to mitigation in the early years of the negotiations. For a long time, adaptation has been the 'poor cousin' of mitigation, almost a 'dirty word' (Khan and Roberts 2013, 174; Burton 1994; Pielke 1998). This lower status of adaptation compared to mitigation has various reasons. Adaptation, it was thought, distracted from the main task of the negotiations, namely to reduce greenhouse gas emissions and avoid the need for adaptation. Adaptation from this perspective was a sign of resignation, a 'defeatist' option accepting that the objective of the Convention—to avoid dangerous climate change—would not be met (e.g. Ciplet et al. 2013; Parry et al. 1998; Pielke 1998; Schipper 2006). Furthermore, discussions of adaptation and adaptation finance were implicitly linked to discussions of responsibility and accountability. Industrialised countries feared that accepting adaptation finance commitments would mean acknowledgement of their responsibility of causing climate change in the first place. Adaptation finance was thus linked to questions of liability and compensation for harm caused—which developed countries wanted to avoid (Ciplet et al. 2013; Gupta 1997; Khan and Roberts 2013; Pielke 1998; Schipper 2006). Finally, developing countries were also internally divided, as Saudi Arabia and other oil-producing states insisted that they also needed support with adaptation, which they understood as diversifying their economy (e.g. Ciplet et al. 2013).

Yet, it increasingly became clear that climate change was occurring and that certain impacts would be felt even if the Kyoto targets were met. And the latter became unlikely, not least because in early 2001, the USA, then the world's largest greenhouse gas emitter, withdrew from the Kyoto Protocol and its binding emissions reduction targets. The focus on mitigation would not be enough and adaptation alongside mitigation was necessary (Burton et al. 2002; Schipper 2006). Discussions about adaptation were linked to questions of finance as well as technology transfer and capacity building. Adaptation, in other words, was a developing country issue (Schipper 2006, 90; Ayers and Dodman 2010).

2.1.2 Adaptation to the Fore: Adaptation Finance from 2001 to 2009

Parties recognised the need for additional, predictable, and adequate funding for non-Annex I countries at COP7, held in Marrakesh in 2001 (UNFCCC 2001, Decision 7/CP.7). The so-called Marrakesh Accords established three new funds: the Special Climate Change Fund (SCCF) and the Least Developed Country Fund (LDCF) under the Convention (UNFCCC 2001, Decision 7/CP.7) as well as the Adaptation Fund under the Kyoto Protocol (UNFCCC 2001, Decision 10/CP.7). While all three funds depend on voluntary contributions from Annex I and Annex II countries, the Adaptation Fund further receives a share of the proceeds of the Clean Development Mechanism, in line with the Kyoto Protocol (UNFCCC 1997, Article 12.8). Adaptation is an explicit objective in all three funds. The aim of the Special Climate Change Fund is 'to finance activities, programmes and measures' in several areas, including climate change.[1] The Least Developed Country Fund aims 'to support a work programme for the least developed countries' which includes '*inter alia*, national adaptation programmes of action' (NAPAs) (UNFCCC 2001, Decision 7/CP.7, para. 6). Finally, the objective of the Adaptation Fund is 'to finance concrete adaptation projects and programmes in developing country Parties that are Parties to the [Kyoto] Protocol' (UNFCCC 2001, Decision 10/CP.7, para. 1).

Although the three funds only became operational after several years (see Gupta 2010, 115ff for an overview), and although they suffered and continue to suffer from insufficient funding, these decisions show how adaptation and adaptation finance arose on the climate change agenda (Gupta 2010; Khan and Roberts 2013). Partly, this rise resulted from a change in the negotiation strategy of developing countries, for whom adaptation and finance were more 'winnable' strategies than mitigation commitments from developed countries (Ciplet et al. 2013; Khan and Roberts 2013). Nonetheless, it remained difficult to turn commitments into concrete action (e.g. Huq 2016).

While efforts to develop a separate 'Adaptation Protocol' on a par with the Kyoto Protocol were unsuccessful (Khan and Roberts 2013; Schipper 2006), Parties made a number of decisions on adaptation. In 2003, adaptation was put on the agenda of the Subsidiary Body for Scientific and Technological Advice (UNFCCC 2003, Decision 10/CP.9). In 2004, Parties adopted the Buenos Aires programme of work on adaptation and response measures (UNFCCC 2004, Decision 1/CP.10)—a 'considerable breakthrough' (Schipper 2006, 89).

The 2007 Bali Action Plan decided on enhanced action in four areas: mitigation, adaptation, technology transfer, and finance (UNFCCC 2007, Decision 1/CP.13). Adaptation—as well as finance—thus had a similar status to mitigation. For enhanced action on adaptation, Parties should

> tak[e] into account the urgent and immediate needs of developing countries that are particularly vulnerable to the adverse effects of climate change, especially the least developed countries and small island developing States, and further taking into account the needs of countries in Africa affected by drought, desertification and floods (UNFCCC 2007, Decision 1/CP.13, para. 1(c)).

The Bali Action Plan should have resulted in a follow-up to the Kyoto Protocol at COP15 in Copenhagen two years later. The Copenhagen Summit, however, failed to deliver such a comprehensive legal agreement. Instead, it resulted in the Copenhagen Accord, with unclear legal status. Because of opposition from a few countries, Parties in the end did not adopt, but only took 'note of', the Accord (UNFCCC 2009, Decision 2/CP.15).[2]

2.1.3 Finance Breakthrough: Adaptation Finance Since 2009

While the Copenhagen Summit is considered a failure in many respects, it did represent a breakthrough on finance for adaptation and mitigation.

In the Copenhagen Accord, Parties recognised that 'enhanced action and international cooperation on adaptation is urgently required' and 'agree[d] that developed countries shall provide adequate, predictable and sustainable financial resources, technology and capacity-building to support the implementation of adaptation action in developing countries' (UNFCCC 2009, Decision 2/CP.15, para. 3). For both adaptation and mitigation, 'scaled up, new and additional, predictable and adequate funding as well as improved access shall be provided to developing countries'. Specifically, Parties committed to 'provid[ing] new and additional resources [...] approaching USD 30 billion for the period 2010–2012 with balanced allocation between adaptation and mitigation'—the so-called 'fast-start finance'—as well as to 'a goal of mobilizing jointly USD 100 billion dollars a year by 2020 [...] from a wide variety of sources, public and private, bilateral and multilateral, including alternative sources of finance'. A 'significant portion' of this funding should be channeled through the newly established Green Climate Fund (UNFCCC 2009, Decision 2/CP.15, para. 8).

While many questions remain unresolved (see below), the $100 billion target has become an important symbolic figure. The Cancún COP formalised the Copenhagen finance commitments, including the $100 billion target, and operationalised the Green Climate Fund (UNFCCC 2010, Decision 1/CP.16, section IV.A). Adaptation received considerable attention. Parties thus affirmed that '[a]daptation must be addressed with the same priority as mitigation' (UNFCCC 2010, Decision 1/CP.16, section I.2) and established an Adaptation Framework as well as an Adaptation Committee to enhance action (UNFCCC 2010, Decision 1/CP.16, section II). Last but not least, Parties began discussions on loss and damage (UNFCCC 2010, Decision 1/CP.16, section II.25 and II.26).

The 2011 Durban COP introduced national adaptation plans (UNFCCC 2011, Decision 5/CP.17), established a Standing Committee on Finance, and undertook a work programme on long-term finance and ways to reach the $100 billion target (UNFCCC 2011, Decision 2/CP.17, section IV), among other things. The following COPs in Doha, Warsaw, and Lima operationalised and clarified many of these provisions, preparing the way for the landmark Paris COP in 2015.

The Paris Agreement that was adopted at the Paris COP was a success in many ways. It establishes for the first time a 'global goal on adaptation of enhancing adaptive capacity, strengthening resilience and reducing vulnerability to climate change' (UNFCCC 2015, Article 7.1). This goal is, however, rather vague, and the Paris Agreement overall is fairly weak on concrete adaptation and finance commitments (see e.g. Roberts and Weikmans 2015; Sharma 2017). It confirms the $100 billion target and requires Parties to 'set a new collective quantified goal from a floor of USD 100 billion per year' before 2025 in its Preamble (UNFCCC 2015, para. 53). The Preamble, which has weaker legal standing than the actual agreement, also urges developed country Parties to develop a 'concrete roadmap to achieve the goal of jointly providing USD 100 billion annually by 2020' and specifically calls for 'significantly increasing adaptation finance' (UNFCCC 2015, Preamble, para. 114). The words 'new and additional' disappeared from the Paris Agreement and Preamble, as did the reference to 'alternative' or 'innovative' sources of funding such as taxes or levies (see also Roberts and Weikmans 2015; Sharma 2017).

The Paris Agreement entered into force on 4 November 2016, less than 1 year after being signed at the Paris COP and just days before the 2016 Marrakesh COP. This represents a record: 'Never before have so many countries joined an international agreement in such a short time' (Fuhr et al. 2016)—an important political signal. Climate finance remained a central and contested agenda item at the Marrakesh summit. On long-term finance, the COP again called on developed country Parties to scale up their climate finance, in particular for adaptation, with a view to achieving a 'greater balance between finance for mitigation and for adaptation' (UNFCCC 2016, Decision 7/CP.22).

2.2 Adaptation Finance or Adaptation Aid? The Question of Additionality

As the previous sections have shown, adaptation and adaptation finance have become more and more important over the course of the climate change negotiations, culminating in the $100 billion target of the 2009 Copenhagen Accord as well as the global adaptation goal of the 2015 Paris Agreement. Yet, many questions remain unresolved. What counts towards the $100 billion? What does a 'balance' between adaptation and mitigation mean? What about accounting, transparency, and additionality? How can we define, report, and monitor adaptation finance (e.g. Roberts and Weikmans 2015, 2017; Sharma 2017)? Here, we focus on the question of additionality and the related question of how climate or adaptation *finance* and adaptation *aid* are similar or different (see Bird 2011). This will help us to obtain a clearer picture of what we are actually analysing in this book.

In principle, climate change represents an additional burden that requires additional resources—resources that are additional to 'traditional' development aid. Accordingly, the Copenhagen Accord as well as the Cancún Agreements—whose legal status is clearer than that of the Copenhagen Accord—stipulate that climate finance be 'new and additional'—yet the texts do not specify a baseline, that is, they fail to clarify *to what* climate finance should be new and additional (see Weikmans 2016b for a discussion).

There have been different interpretations of what constitutes 'new and additional' resources (Brown et al. 2010; Klein 2010; Stadelmann et al. 2010). Many developing countries would like to see a complete separation between development aid and climate finance flows because of the different motivations behind these two flows. Climate finance, as opposed to development aid, is not voluntary or charity, but an obligation or a form of compensation, given the disproportionate responsibility of developed countries for causing climate change (Duus-Otterström 2015; Huhtala et al. 2010; Klein 2010; Oxfam International 2007; Weikmans 2016a,b; see also Chap. 6). Developed countries, as we have seen in Sect. 2.1, reject discussions of liability and compensation. For them, climate finance is part of a deal of getting developing countries to sign up to mitigation commitments (Weikmans 2016b). Accordingly, many tend to interpret climate finance additionality as aid over and above the 0.7% aid target. Donors have pledged to provide 0.7% of their GNI as official development assistance; any aid above that percentage would be new and additional (Ayers and Abeysinghe 2013; Brown et al. 2010). This baseline, however, is problematic as few donors have (as yet) met the 0.7% commitment. The UK and Sweden both meet the 0.7% target. The former even enshrined this target into law, while the latter formally committed to providing 1% of its GNI to aid (Government of Sweden 2014; Parliament of the United Kingdom 2015; see also Chap. 6). Some countries—most importantly Germany—have instead taken 2009 as a baseline and interpreted 'new and additional' to mean above what they spent on climate aid in 2009. Other countries proposed to specify a percentage of aid going to climate actions (see Brown et al. 2010, 2).

Without a clear baseline, it is hard to assess to what extent climate and adaptation finance[3] is indeed new and additional. Accordingly, developing countries and some observers have repeatedly voiced concerns that the pledged adaptation finance is not 'new and additional', but that industrialised countries simply rebrand their development aid as relevant for adaptation (or mitigation)—at the expense of other, equally important, development challenges (e.g. Ciplet et al. 2013, 59; Adaptation Watch 2015; Ayers and Abeysinghe 2013; Carty et al. 2016; Dasgupta and Climate Change Finance Unit 2015; Klein 2010). Developed countries, on the other hand, emphasise the links and synergies between development and adaptation. From this perspective, separating adaptation and development finance risks a duplication of efforts and misallocation of scarce resources (Smith et al. 2011, 988).

The distinction between aid for development and aid for adaptation is indeed a difficult one, since adaptation and development are linked in different ways. Social and economic welfare and development underpin adaptive capacity and therefore vulnerability (see Chap. 1; Ayers and Abeysinghe 2013, 489; Ayers and Dodman 2010; Fankhauser 2010; Fankhauser and Burton 2011; Weikmans 2016b). Development, in this sense, is thus often synonymous with adaptation, and adaptation measures and development measures can reinforce each other:

> Good (or sustainable) development (policies and practice) can (and often does) lead to building adaptive capacity. Doing adaptation to climate change often also means doing good (or sustainable) development (Huq and Ayers 2008, 52, cited in Ayers and Dodman 2010, 165).

Similarly, climate change threatens development progress. Without adaptation measures, past development achievements are at risk (Weikmans 2016b; see also Chap. 6). In this context, Diamond and Bruch (2011) estimate that 60% of all development assistance could intersect with adaptation activities, while a 2006 World Bank report estimates that climate change may threaten 40% of its total portfolio (World Bank 2006, 27). Analysis by the United Nations Development Programme (UNDP) suggests that one-third of all aid flows are affected by climate change, of which half (or 17% of total aid flows) is highly sensitive to climate risks (UNDP 2007, 190f).

Given the links between adaptation and development, 'it makes sense to think of adaptation not as an incremental activity to deal with climate change, but as climate-resilient development' (Fankhauser 2010, 24) or as 'development in a hostile climate' (Stern 2009, cited in Fankhauser 2010, 24). To some extent, adaptation finance reflects this understanding of adaptation; most adaptation finance to date has come as ODA or development aid: 'almost all funding for adaptation currently comes from public finance, drawn from international aid budgets' (Ayers and Abeysinghe 2013, 494;

see also Weikmans 2016a). This raises of course criticisms regarding the additionality of resources, as discussed in this section. We do not seek to solve this issue here; instead, we recognise that donors largely support adaptation through their bilateral and to a lesser extent multilateral aid. Hence, the analysis in this book focuses on bilateral aid flows marked as relevant for adaptation—although we do acknowledge that adaptation aid is conceptually different from adaptation finance and that aid flows as registered by the OECD are contested (see also Chaps. 3 and 7).

2.3 Aid Allocation

Now that we have an overview of the ascent of adaptation and adaptation finance in the climate change negotiations, we can turn to the central topic of this book: adaptation aid allocation. Since adaptation aid is a subset of overall development aid, the literature on how donors distribute their *development aid* may give us some indication of how they allocate their *adaptation aid*. In this and the following section, we therefore introduce theories and models of aid allocation in general, and with regard to adaptation aid specifically, and we derive hypotheses to be tested in later chapters.

Let us first look at the allocation of development aid in general. Development aid is defined as a transfer of concessional finance for the promotion of economic development and welfare to countries and territories on the list of ODA recipients (OECD 2016a; see Chap. 1). The OECD Development Assistance Committee (DAC) publishes this list of ODA recipients and revises it every 3 years, based on GNI per capita.[4] In 2015, bilateral aid from all DAC donors amounted to about $131 billion (OECD 2016b). This puts the $100 billion target for climate finance mentioned earlier in perspective and indicates just how significant climate finance is.

Where does all this aid go? When allocating aid, a donor needs to make several decisions. Through which channels and modalities should the aid be provided? Which countries should the donor support, and what programmes or projects should be funded within the selected countries? We focus on bilateral aid, as does much of the literature, not only because bilateral aid flows are much larger than multilateral flows, but also because bilateral aid leaves donors with much more leverage as to how they allocate their aid (Gulrajani 2016; see also Chap. 6). Our focus is on the question of geographical distribution of bilateral aid: which countries do donors support in their adaptation efforts—and why? This question of aid allocation across countries has received considerable attention since the 1970s. A large body of literature has examined development aid more broadly, as well as specific types of aid such as humanitarian aid or 'green' aid, aid with an environmental focus. The broader development aid literature has identified three factors or models of aid allocation: recipient need, recipient merit, and donor interests. In practice, donors seem to take into account all three factors, though their own interests often seem to matter the most. A much smaller literature has tested the role of these factors specifically for the allocation of adaptation aid (see Sect. 2.4). Here, empirical findings indicate that donors pay little attention to recipient need, which is understood as vulnerability to climate change in an adaptation context.

2.3.1 *Recipient Need vs Donor Interests*

In the 1970s, studies discussed the motivations of donors to give aid and distinguished between two main ones or utilities of aid: charitable motives, that is, 'the simple desire to help the less fortunate' (Dudley and Montmarquette 1976, 132) on the one hand, and strategic and commercial considerations on the other (Dudley and Montmarquette 1976; McKinlay 1978; McKinlay and Little 1977, 1979). The former British Prime Minister David Cameron described these two motivations as the 'heart' versus 'head' arguments (cited in Lightfoot et al. 2016). Accordingly, a donor may maximise the economic assistance utility and allocate aid proportional to recipients' development needs—what McKinlay and Little (1977) refer to as the *recipient need* model and David Cameron as the 'heart' argument. Alternatively, a donor may maximise the foreign policy utility of aid and provide it largely to pursue its own commercial, political, or security interests—what McKinlay and Little (1977) refer to as the *donor interest* model and David Cameron as the 'head' argument (Lightfoot et al. 2016). These two models are, however, not mutually exclusive; aid 'potentially contains both economic assistance and foreign policy utilities' (McKinlay 1978, 240).

If donors follow their heart, that is, the recipient need model, we should see a negative relationship between the level of socio-economic welfare and development and the level of aid per capita. The less developed a country, the higher its development needs and the more aid it should receive. Need is an attribute of the recipient; assuming that donors understand need in a similar way—as poverty—we therefore should see a similar allocation pattern across different donors.

Mostly, recipient need is measured by per capita income (e.g. Alesina and Dollar 2000; Berthélemy 2006a; Clist 2011; Hoeffler and Outram 2011; Younas 2008). Average per capita income, however, hides income inequalities that may be very high in some recipient countries (see Younas 2008, 665). Several authors have thus complemented per capita income with additional measures. McKinlay (1978) for instance operationalise recipient need with seven different indicators; beyond GDP per capita, they also use for example calorie consumption per capita or the number of doctors per 100,000 inhabitants. McKinlay and Little (1977) further include recipients' performance capabilities as measured by variables like gross domestic capital formation or size of the manufacturing or mining sector. Other studies have used additional measures of economic performance, such as indebtedness (Berthélemy 2006a,b)[5] or social outcome variables, such as infant mortality rates (Younas 2008) or the percentage of the population living below the national poverty line or on less than $2 a day (Figaj 2010).

If donors in contrast follow their head, that is, the donor interest model, we should see a positive relationship between the foreign policy relevance of a country to a specific donor and the level of aid it receives from that specific donor. The more important—from an economic, political, or security point of view—a country is to a specific donor, the more aid it should receive from that donor. As opposed to recipient need, donor interests are specific to a donor–recipient pair, and we should therefore see different allocation patterns across donors. A recipient country may for instance trade a lot with donor *A* but not very much with donor *B*; we would expect this country to receive much higher levels of aid from donor *A* than from donor *B*.

Again, donor interests have been operationalised in different ways, since donors have different interests. The literature distinguishes mainly between commercial or economic interests on the one hand and geopolitical interests on the other. Some authors add security interests as a third type of interest.

Measures of commercial interests relate mainly to trade, notably the flow of exports from a donor to a given recipient (e.g. Alesina and Dollar 2000; Berthélemy 2006b; Clist 2011; Dollar and Levin 2006). Some studies (also) use other measures, including: imports in percent of a recipient's GDP, either total or for specific sectors (Younas 2008); the total level of foreign direct investments (Alesina and Dollar 2000; Halimanjaya 2015); or trade openness (Alesina and Dollar 2000).

A number of variables measure geopolitical interests or proximity, with the expectation that donors favour countries that are closer to them—geographically, culturally, or politically. Former colonial status is one common measure of (political) proximity (e.g. Berthélemy 2006a,b; Clist 2011; Dollar and Levin 2006; Dudley and Montmarquette 1976; Younas 2008). Other measures are geographic proximity (e.g. Clist 2011; Dollar and Levin 2006; Hoeffler and Outram 2011) or a common language (Clist 2011) or religion (Alesina and Dollar 2000; Clist 2011; Younas 2008) for cultural proximity. Another measure of political proximity is voting similarity in the United Nations General Assembly, that is, how often a recipient and a donor vote the same way (Alesina and Dollar 2000; Dreher et al. 2008; Hoeffler and Outram 2011). Dreher et al. (2008) for instance find that voting in line with the USA can increase development aid flows from that country. Drury et al. (2005) instead rely on international alliance data to capture political proximity. Finally, security interests seem to be of particular importance to the USA, which gives disproportionate amounts of aid to Israel and Egypt because of these countries' political roles in the Middle East. Several studies hence explicitly control for Israel and Egypt (Alesina and Dollar 2000; Younas 2008). Clist (2011) uses the total arms exports of a donor to a recipient—though this probably measures economic rather than security interests.

A large number of studies have tested the role of recipient need versus donor interests. It is hard to compare the specific findings of these studies, not only because they have operationalised recipient need and donor interests in different ways, as we have seen. They also cover different time periods, different donors (both multilateral and bilateral), and different aid flows, for instance, humanitarian aid (e.g. Drury et al. 2005; Kevlihan et al. 2014), environmental aid (Hicks et al. 2008), or aid for climate change mitigation (Halimanjaya 2015). Nonetheless, some common findings emerge. Most studies find evidence for both motivations, that is, both recipient needs and donor interests influence how donors allocate their aid, though the relative role of these two motivations differs. In general, it seems that donor interests dominate, in particular commercial interests, as already suggested in the 1970s (Dudley and Montmarquette 1976; McKinlay 1978; see also e.g. Alesina and Dollar 2000; Berthélemy 2006a; Drury et al. 2005). Berthélemy (2006a, 88) thus concludes that 'bilateral aid motives are, to a large extent, egoistic rather than altruistic'. Alesina and Dollar (2000, 33) similarly find 'considerable evidence that the pattern of aid giving is dictated by political and strategic considerations'.

However, when looking at individual donors, we get a more nuanced picture. There are striking differences across donors with regard to how much weight they give to recipient need compared to donor interests (e.g. Berthélemy 2006a,b; Clist 2011). While some donors tend to focus more on poverty reduction (that is, recipient need), others are much more guided by their own economic, political, and/or security interests. The Nordic countries are known for their altruism; they not only are more generous regarding how much aid they give, but also give most of their aid to poor countries (as well as those with democratic institutions, see below) (Clist 2011). The Netherlands and Switzerland show similar tendencies (Berthélemy 2006a,b; Clist 2011). Other countries are more egoistic and pay greater attention to their own interests, notably economic ones (e.g. Berthélemy 2006b). France tends to favour its former colonies (e.g. Alesina and Dollar 2000), while the USA is mainly guided by its security interests in the Middle East—as the large amounts of aid to Israel and Egypt mentioned earlier indicate (e.g. Alesina and Dollar 2000; Younas 2008).

2.3.2 Recipient Merit

In the early 2000s, a third motivation was added to the basic distinction between recipient need and donor interests: recipient merit. There are two arguments for why recipient countries with 'good' policies and governance merit more adaptation aid: effectiveness and the intrinsic value of good governance. On the one hand, aid is arguably more effective in countries with sound economic policies and stable political institutions, as the 2002 Monterrey Consensus on Financing for Development[6] explicitly acknowledges:

> Sound economic policies, solid democratic institutions responsive to the needs of the people and improved infrastructure are the basis for sustained economic growth, poverty eradication and employment creation. [...] Sound policies and good governance at all levels are necessary to ensure ODA effectiveness (United Nations 2002, 6; 14).

On the other hand, donors intrinsically value good economic and political governance. By allocating aid to countries with good governance, they not only increase the effectiveness of their aid but also reward and promote democratic or democratising institutions and good governance (e.g. Younas 2008; Zanger 2000).

Just like recipient needs and donor interests, recipient merit has been operationalised in different ways, although different measures of democracy dominate. Many authors opt for data from Freedom House that measures the extent of civil liberties and political rights (e.g. Alesina and Dollar 2000; Clist 2011; Dollar and Levin 2006; Younas 2008). To a lesser extent, studies use another democracy measure such as the Polity IV data (Hoeffler and Outram 2011). These democracy measures capture political institutions; others have also taken into account economic institutions through measures of the rule of law (Dollar and Levin 2006) or per capita GDP growth (Hoeffler and Outram 2011). Another operationalision of recipient merit focuses on the extent of conflict; studies here have used a dummy for the presence of interstate or internal conflict (Berthélemy 2006a) or the level of political terror (Clist 2011).

Overall, empirical results suggest that recipient merit matters. Countries that are more democratic and respect political and human rights as well as the rule of law receive—on average—more foreign aid (Dollar and Levin 2006; Hoeffler and Outram 2011; Younas 2008). Figaj (2010), in contrast, finds no significant effect of freedom (her measure of democracy) and the level of environmental aid countries receive. For economic governance, measured by the rule of law, Dollar and Levin (2006) find a positive effect in their more recent period of analysis (2000 through 2003), though this is only significant in the case of multilateral aid, and a negative and significant effect for their earlier period of analysis (1984 through 1989) for both bilateral and multilateral aid. The authors thus conclude that 'aid used to be targeted to countries with poor economic governance' (Dollar and Levin 2006, 2036).

2.3.3 *Other Factors*

There are three additional factors that may influence how donors allocate their aid: population, path dependency or what Barrett (2014) calls 'donor utility', and network effects.

Population has generally a large and significant effect on aid allocation. Small countries—in terms of their population—receive on average more aid per capita than more populous countries. This is the well-documented 'small country bias' or 'small country effect' (e.g. Alesina and Dollar 2000; Clist 2011; Dudley and Montmarquette 1976; Younas 2008). The literature suggests three explanations for this effect. First, the marginal impact of aid decreases with increasing population. Second, smaller countries have higher administrative capacities and can better absorb aid. And third, donors can more easily influence smaller countries compared to larger countries (Younas 2008, 667).

Path dependency refers to the effect of past aid allocation decisions on current allocation decisions. Recipient countries in which a given donor has been active in the past are more likely to benefit from future support from that donor. An existing aid relationship and established networks reduce transaction costs and increase aid effectiveness (Barrett 2014; Robertsen et al. 2015).

Finally, donors do not make their aid allocation decisions in isolation, but take into account the decisions of other donors. This may go in two opposite directions: coordination or herding. In the first case of coordination, donors follow a 'division of labour'. If donor *A* provides aid to a given recipient *R*, donor *B* will instead focus its aid on recipient *S*. In the latter case of herding, donors instead provide aid to similar countries simultaneously. Accordingly, if recipient *R* already receives aid from donor *A*, it is quite likely that donor *B* also provides aid to this recipient. Similarly, if a recipient receives only little or no support from donor *A*, it is unlikely that donor *B* will provide aid to this recipient. Although few studies have explicitly tested the interactions between donors' allocation decisions, herding seems to occur in practice. This is the story of so-called aid 'darlings' and 'orphans' (Berthélemy 2006b; Davies and Klasen 2013; Hoeffler and Outram 2011; Klasen and Davies 2011). Hoeffler and Outram (2011) control for the amount of aid per capita a recipient receives from all other bilateral donors, but understand this as a measure of recipient merit, because high levels of aid from other donors serves as a signal of good governance and high aid effectiveness. Berthélemy (2006b) controls for the amount of aid a recipient receives from other bilateral donors as well as from multilateral donors; he sees only the second variable as a measure of recipient merit.420660

2.4 Allocating Adaptation Aid

Now that we have an overview of the general development aid literature, and the expectations and findings of development aid allocation, we can turn to the allocation of adaptation aid specifically. How do the different determinants presented in the previous section play out in the case of adaptation aid? A large body of literature has tested the role of recipient need, donor interests, and recipient merit for specific aid flows, including 'green' aid, that is, aid with an environmental dimension (Figaj 2010; Hicks et al. 2008; Lewis 2003; Miller 2014). Increasingly, studies have also specifically analysed the allocation of climate-related aid, for mitigation (Halimanjaya 2013, 2015) as well as for adaptation. In the context of environmental aid and adaptation aid, recipient need is no longer understood by poverty alone. As we discussed earlier, there has been international agreement in the climate change negotiations that adaptation finance should focus on 'particularly vulnerable' countries. We turn to this discussion again and then focus on the difficulties of identifying which countries are particularly vulnerable, that is, of measuring vulnerability.

2.4.1 Vulnerability as Recipient Need

Development aid, by definition, aims to promote economic development and welfare (OECD 2016a). Adaptation aid further seeks to 'reduce the vulnerability of human or natural systems to the impacts of climate change and climate-related risks', according to the OECD guidelines which we use to identify adaptation relevant aid (OECD 2011, 4; see also Chap. 3).

Following this definition, recipient need, in an adaptation context, is broader than just poverty and also includes vulnerability to the adverse effects of climate change. Countries that are more vulnerable have a greater need of adaptation and hence also of adaptation aid. As we have seen in the introduction (Chap. 1), vulnerability is composed of exposure and sensitivity to climate risks as well as adaptive capacity. Countries that are more exposed and more sensitive, for instance low-lying countries or drought-prone countries, are more vulnerable, as are countries that have less adaptive capacity. Adaptive capacity is related to poverty, since poverty is a direct driver of adaptive capacity. The fewer resources a country has, the lower is its capacity to adapt. Adaptive capacity further depends on institutional and political structures (Smit and Wandel 2006; Turner et al. 2003; see Chap. 1). Countries with democratic institutions, good governance, and sound (environmental) policies tend to be more able to deal with the consequences of climate change and therefore are less vulnerable. As a consequence, in the context of adaptation aid, recipient need and recipient merit are closely connected (see also Chap. 3).

There is broad agreement that adaptation finance should be given based on recipient need, that is, that the most or particularly vulnerable countries be put first. We have seen that international agreements since the 1992 Convention have repeated the focus on 'particularly vulnerable' developing countries (see Sect. 2.1). Several academics similarly argue for the prioritisation of particularly or most vulnerable countries from a justice perspective (Duus-Otterström 2015; Grasso 2010a,b; Klein and Möhner 2011; Weikmans 2016a).

At the same time, donors seek to maximise the impact of their support and thus focus their aid to countries where its (potential) impact is greatest. Accordingly, in the 'Roadmap to US$100 Billion', the OECD (2016c, 19) states that 'without accessible and catalytic finance that flows to where it is most needed and has the greatest impact, any quantity of finance will fall short of the Paris Agreement goals'. In a similar vein, Barr et al. (2010, 845) highlight the importance of implementation capacity—'the ability to manage and use finance effectively'—as a criterion for adaptation aid allocation alongside vulnerability, given that adaptation finance will be scarce and thus needs to be spent efficiently and effectively.

There are thus two principles that potentially guide adaptation finance allocation: equity on the one hand and cost-effectiveness or efficiency on the other. An equitable distribution of resources would prioritise vulnerable countries whereas a cost-effective or efficient distribution would prioritise countries where net benefits of adaptation aid are largest (Stadelmann et al. 2014; Weikmans 2016a). To some extent, these principles are mutually exclusive, as countries that are highly vulnerable because of their low adaptive capacity also tend to struggle with using aid effectively—a point we will return to in Chap. 3. At least on paper, equity seems to be the dominant principle, as the focus on 'particularly vulnerable' countries in the negotiation texts indicate (see also e.g. Klein and Möhner 2011; Roberts et al. 2017; Weikmans 2016a). For some, the dominance of equity results from the difficulties of assessing the cost and benefits of adaptation actions (Persson and Remling 2014; Weikmans 2016a; see also Sect. 1.2)—though measuring vulnerability is no easy task either, as the definition of vulnerability in the introduction has already hinted at.

2.4.2 *Measuring Vulnerability*

If recipient need and recipient merit are closely linked and hard to separate, as the discussion in the previous section indicates, we need to take a closer look at how vulnerable countries are identified, that is, how climate change vulnerability is measured.

Policy-makers and climate negotiators have repeatedly called on science to develop generic vulnerability indices (Füssel 2010; Klein 2009). Researchers have responded to the demand for such indices and a range of different quantitative indices exist (e.g. DARA 2012; Guillaumont and Simonet 2011; Kaly et al. 2004; ND-GAIN n.d.; Wheeler 2011; see also Chap. 3). Yet, this has—unsurprisingly—not led to 'a systematic and agreed way of assessing, measuring, expressing and comparing the vulnerability of countries to climate change' (Klein and Möhner 2011, 16). Reviews of different vulnerability indices point to 'fundamental conceptual, methodological, and/or empirical flaws' (Füssel 2010, 598). Problems relate in particular to selecting and aggregating appropriate proxies on the one hand, and dealing with the time-specific and place-specific nature of vulnerability on the other.

Vulnerability is composed of exposure, sensitivity, and adaptive capacity (see Chap. 1). How can we measure these three components, and how should these measures be aggregated into one vulnerability index? Mostly, vulnerability indices separate physical measures of exposure and sensitivity from socio-economic measures of adaptive capacity (Hinkel 2011; Muccione et al. 2017), though many indices capture only or largely the former aspect of vulnerability, that is, physical exposure and sensitivity (see Eriksen and Kelly 2007). The indices use a wide variety of proxies. Brooks et al. (2005, 155) for instance identify 46 different 'potential proxies for national-level vulnerability to climate change', including GDP per capita, the percentage of the population living within 100 km of the coastline, or the percentage of land area covered by forest.

The indices also use different methods to aggregate the selected variables into one index; these methods are not always clear or justified (Eriksen and Kelly 2007; Füssel 2009; Weikmans 2016a). The selection of variables is constrained by data availability and inevitably involves normative choices, as does the aggregation of variables into one national-level index (Füssel 2009, 2010; Hinkel 2011).

That most vulnerability indices focus on the national level relates to the second key problem: the level at which vulnerability should or could be measured. Most indices assess vulnerability at the national level, not least because this is the level for which allocation decisions for adaptation finance are largely made (Eriksen and Kelly 2007, 507; Muccione et al. 2017). Yet, the national level may not be the most appropriate one for assessing vulnerability, which is place-specific and system-specific (Smit and Wandel 2006, 286f) and varies not only across regions, but also across and even within communities (e.g. Eriksen and Kelly 2007). A low-lying coastal community is exposed and sensitive to different risks than a community in a flood-prone mountainous area in the same country. Similarly, the most marginalised and poorest members of a community are most vulnerable to climate change, not only because they tend to be more physically exposed and sensitive but also because they have less adaptive capacity. Think of the poor subsistence farmer in an arid region who depends on rain-fed agriculture and has no means to diversify his livelihood or purchase insurance against drought (see Chap. 1). An indicator at the national level hides these differences in vulnerability across regions, communities, and groups of society.

There are additional problems of measuring vulnerability, for instance the 'forward-looking aspect of vulnerability': vulnerability to climate change is about potential harm from future climate impacts, which are uncertain (Hinkel 2011). Given these different difficulties, some authors conclude that it is impossible to quantify vulnerability, at least at the national level: 'vulnerability, like happiness, is a human state or condition that cannot be measured directly in any objective fashion', write Eriksen and Kelly (2007, 500), a conclusion others agree with (Füssel 2009; Hinkel 2011; Klein 2009; Moss et al. 2001). In this context, Hinkel (2011, 200) argues that speaking of vulnerability 'measurement' is misleading; 'operationalisation' of vulnerability is more accurate since the latter is a theoretical concept. We agree that measuring or operationalising vulnerability is inherently difficult, and that 'allocating funds based on the assessment of vulnerability is a process fraught with ambiguity' (Ciplet et al. 2013, 60). We seek to circumvent this problem by using a range of vulnerability indicators, as we describe in greater detail in Chap. 3. Other empirical studies have similarly used different vulnerability indicators. We now turn to these studies to assess to what extent vulnerability—measured in different ways—seems to matter in practice for adaptation aid allocation.

2.4.3 *Does Vulnerability Matter?*

Several studies have empirically examined whether vulnerable countries receive more adaptation aid, for specific donors and/or recipients, across all donors, at the sub-national level, or for multilateral adaptation aid. Generally, these studies have applied the recipient need, recipient merit, and donor interest models of aid allocation to the case of adaptation aid flows, and used different vulnerability indices to operationalise recipient need. The general conclusion is that vulnerability plays at best a small role: recipient merit and, to a lesser extent, donor interests seem to explain better the distribution of adaptation aid.

Barrett (2014) examines sub-national adaptation aid allocation across districts in Malawi. His analysis includes both an indicator of physical vulnerability—a measure of drought and dry spells as well as floods—and two indicators of socio-economic vulnerability—infant mortality and life expectancy. The results show that more aid-funded adaptation projects are located in physically vulnerable districts, while the opposite is true for socio-economically vulnerable districts: the analysis indicates a negative relationship between infant mortality and life expectancy on the one hand, and adaptation aid on the other. The author thus concludes that 'the poorest, most marginalized, and climate vulnerable districts receive the least adaptation finance within Malawi' (Barrett 2014, 131; see also Barrett 2013).

Robertsen et al. (2015) examine adapation aid flows from seven donors to sub-Saharan Africa using the exposure sub-index of the Notre Dame Global Adaptation Index (ND-GAIN) and GDP per capita as measures of vulnerability. Neither variable seems to drive adaptation aid flows. Robinson and Dornan (2017) focus mainly on adaptation aid in SIDS; they use the exposure and the sensitivity sub-indices of the ND-GAIN, losses from weather extremes, and GDP per capita to measure vulnerability. While they do find that poorer countries receive more adaptation aid, they do not find a consistent relationship between physical vulnerability and adaptation aid. Betzold (2015) examines German adaptation aid and similarly finds that Germany assists poorer but not more physically vulnerable countries with adaptation. Nakhooda et al. (2013) assess fast-start finance as reported to the UNFCCC, and conclude that fast-start finance is only weakly correlated with the two measures of vulnerability used, the DARA and ND-GAIN indices. In contrast, we conclude from our own analyses of all donors (Betzold and Weiler 2017; Weiler et al.

2017) that countries more exposed to weather-related climate extremes and sea level rise receive more adaptation aid. For GDP per capita, we find a nonlinear relationship: the poorest and richest countries receive less adaptation aid, with countries with a GDP per capita of around $1,000 receiving the highest levels per capita. Several studies also explicitly test whether SIDS, LDCs, or African countries—the countries singled out as 'particularly vulnerable' in the climate change negotiations—receive more adaptation aid. While SIDS indeed receive more, this is not the case for LDCs or African countries (Betzold and Weiler 2017; Robinson and Dornan 2017; Weiler et al. 2017).

What about multilateral funds? Do they prioritise vulnerable countries? At least on paper, vulnerability is important for multilateral adaptation funding. The Green Climate Fund specifically reserves half of its adaptation funding for SIDS, LDCs, and African countries (Green Climate Fund 2014), while the Adaptation Fund finances projects in 'particularly vulnerable' developing country Parties to the Kyoto Protocol (UNFCCC 2008, Annex IV, section III.10; see also Horstmann 2011). Stadelmann et al. (2014) and Remling and Persson (2015) analyse which projects the Adaptation Fund approves. Both studies conclude that neither equity—the level of vulnerability—nor efficiency or cost-effectiveness guide funding decisions. The Adaptation Fund 'has rather approved projects from high-income and less vulnerable countries with high absolute economic savings, while not approving projects in poor, vulnerable countries with high relative economic savings' (Stadelmann et al. 2014, 116). Weikmans (2016a) concludes from his literature review that vulnerability may have played a role in the selection of programmes, projects, or countries, but that it did not determine the level of multilateral funding.

If vulnerability does not drive adaptation aid flows, which factors do? What is the role of donor interests and recipient merit? Recipient merit is an important predictor of adaptation aid. According to most analyses, good governance measures such as the extent of political freedom or control of corruption are strongly related with the level of adaptation aid. The better governed a country, the more adaptation aid it can expect, all else being equal (Betzold 2015; Betzold and Weiler 2017; Robertsen et al. 2015; Robinson and Dornan 2017; Weiler et al. 2017). The evidence for donor interests is more mixed. Our own research suggests that donor interest variables such as trade or voting in the United Nations General Assembly (Betzold 2015; Weiler et al. 2017) influences how much adaptation aid a country receives. Other authors find only a weak link between adaptation aid and donor interests, measured by colonial ties (Robertsen et al. 2015; Robinson and Dornan 2017). However, adaptation aid flows mirror general development aid flows: countries that receive high levels of development aid in general can also expect high levels of adaptation aid (Barrett 2014; Betzold and Weiler 2017; Nakhooda et al. 2013; Robertsen et al. 2015; Weiler et al. 2017). And as discussed earlier, development aid is largely given to promote donors' own foreign policy objectives.

Finally, two additional factors influence adaptation aid flows: population and transaction costs or absorptive capacity. All studies of adaptation aid allocation control for population and find the same small country bias that studies of overall development aid have documented. On the one hand, larger countries—in terms of their population—are more likely to receive some adaptation aid. On the other hand, the smaller a country's population, the higher the level of adaptation aid *per capita* (Betzold and Weiler 2017; Robinson and Dornan 2017; Weiler et al. 2017). Barrett (2014) finds strong evidence for what he terms donor utility: districts with lower transaction costs—because of existing aid networks—and better absorptive capacity receive more adaptation aid. Robertsen et al. (2015) also interpret the strong correlation between overall development aid and adaptation aid as an indicator of transaction costs: donors are more likely to support adaptation activities in countries with which they have an ongoing aid relationship.

2.5 Summary and Our Expectations

We started this chapter with an overview of the climate change negotiations and specifically the evolution of adaptation and adaptation finance in these negotiations. We have seen that adaptation and adaptation finance have gained prominence over time and are at the centre of current negotiation rounds and agreements. Yet, what counts as climate finance, and how this finance should be distributed, are key—and strongly contested—questions in the negotiations. To what extent traditional development assistance represents new and additional finance remains controversial; in practice, however, much climate finance, especially adaptation finance, comes as development aid. This dominance of development aid in the provision of climate finance motivates our empirical focus on adaptation aid, or ODA that is relevant for adaptation.

To understand how adaptation aid is distributed geographically, we therefore turned to research on development aid more broadly. Patterns of aid allocation across recipient countries have been the focus of a large number of studies since the 1970s. This literature has identified three factors or models that explain how donors allocate their development aid: recipient need, recipient merit, and donor interests. Similar factors seem to be at play with regard to adaptation aid, where recipient need means vulnerability to climate change. Although scholars and practitioners agree that vulnerable countries should be prioritised in the allocation of adaptation aid, the empirical evidence suggests that such countries do *not* receive systematically more adaptation aid than other countries. To some extent, however, the rather weak evidence may result from the inherent difficulties of operationalising and measuring vulnerability to climate change.

With this study we wish to contribute to the literature on (adaptation) aid allocation. Building on the current literature, including research on adaptation aid, we would *a priori* expect similar factors to be at play as those we observe for development aid in general. In other words, donors should allocate their adaptation aid according to three factors: recipient need, understood as vulnerability to climate change; recipient merit; and donor interests. For adaptation aid, however, recipient need and recipient merit are closely related, as we have seen earlier: good governance implies better adaptive capacity and thus lower vulnerability. We have thus two opposite expectations for the relationship between good governance and adaptation aid. If we find a *negative* relationship between good governance and adaptation aid, we take this to indicate that donors allocate their aid based on recipient need, for worse governed countries have lower adaptive capacity and hence need more support with adaptation. On the other hand, if we find a *positive* relationship between good governance and adaptation aid, we take this to indicate that donors allocate their aid based on recipient merit, for better governed countries are better able to use funds efficiently and hence merit more support, including with adaptation to climate change. Since we want to test separately how the different components of vulnerability influence adaptation aid flows, we formulate two expectations with regard to recipient need (H1a and H1b). This leaves us with a total of four expectations:

H1a The more exposed and sensitive a country to the adverse effects of climate change, the more adaptation aid it receives.

H1b The lower the adaptive capacity of a country, the more adaptation aid it receives.

H2 The better governed a country, the more adaptation aid it receives.
H3 The more economically or politically relevant a country to a donor, the more adaptation aid it receives from that donor.

As opposed to other studies, we use a mixed methods research design to test these hypotheses. We not only examine adaptation aid flows as reported in the OECD Creditor Reporting System (CRS) using quantitative regression analyses, but also take a closer look at the actual decision-making processes through qualitative case studies of three large climate donors: Germany, Sweden, and the UK. In the next chapter, we describe our methods in greater detail, list our variables for the quantitative analyses, and document the interviews and policy documents used in the qualitative analysis.

Notes

1. The other areas are technology transfer; energy, transport, industry, agriculture, forestry and waste management; and economic diversification (UNFCCC 2001, Decision 7/CP.7, para. 2).
2. For a discussion of the Copenhagen Summit, see for instance Bodansky (2010).
3. In the following, we focus only on adaptation finance. Many of the arguments also apply to mitigation finance and thus to climate finance more broadly.
4. The list can be found at http://www.oecd.org/dac/stats/daclist.htm.
5. Indebtedness could also be a measure of donor interest: 'donors may be locked in a "debt game," in which they have to provide new resources to highly indebted countries simply to avoid that these debtors fall in arrear ' (Berthélemy 2006b, 184).
6. The 'Monterrey Consensus' is the outcome of the United Nations Conference on Financing for Development, held in Monterrey, Mexico, in 2002. See http://www.un.org/esa/ffd/overview/monterrey-conference.html for more information.

References

Adaptation Watch. 2015. Toward Mutual Accountability: The 2015 Adaptation Finance Transparency Gap Report. Policy Briefing. Available online at http://www.adaptationwatch.org/.

AdaptationWatch. (Weikmans et al.) 2016. Towards Transparency. The 2016 Adaptation Finance Transparency Gap Report. White Paper. Available online at http://www.adaptationwatch.org/.

Alesina, A., and D. Dollar. 2000. Who Gives Foreign Aid to Whom and Why? *Journal of Economic Growth* 5: 33–63.

Ayers, J.M., and A.C. Abeysinghe. 2013. International Aid and Adaptation to Climate Change. In *The Handbook of Global Climate and Environmental Policy*, chap. 28, ed. R. Falkner, 486–506. New York: Wiley.

Ayers, J.M., and D. Dodman. 2010. Climate Change Adaptation and Development I: The State of the Debate. *Progress in Development Studies* 10 (2): 161–168.

Barr, R., S. Fankhauser, and K. Hamilton. 2010. Adaptation Investments: A Resource Allocation Framework. *Mitigation and Adaptation Strategies for Global Change* 15: 843–858.

Barrett, S. 2013. Local Level Climate Justice? Adaptation Finance and Vulnerability Reduction. *Global Environmental Change* 23: 1819–1829.

Barrett, S. 2014. Subnational Climate Justice? Adaptation Finance Distribution and Climate Vulnerability. *World Development* 58: 130–142.

Berthélemy, J.-C. 2006a. Aid Allocation: Comparing Donors' Behaviours. *Swedish Economic Policy Review* 13: 75–109.

Berthélemy, J.-C. 2006b. Bilateral Donors' Interest vs. Recipients' Development Motives in Aid Allocation: Do All Donors Behave the Same? *Review of Development Economics* 10 (2): 179–194.

Betzold, C. 2015. Vulnerabilität, Demokratie, politische Interessen? Wie Deutschland seine Hilfe für Anpassung an den Klimawandel verteilt. [Vulnerability, Democracy, Political Interests? How Germany Allocates its Aid for Adaptation to Climate Change]. *Zeitschrift für Internationale Beziehungen* 22 (1): 75–101.

Betzold, C., and F. Weiler. 2017. Allocation of Aid for Adaptation to Climate Change: Do Vulnerable Countries Receive More Support. *International Environmental Agreements: Politics, Law and Economics* 17 (1): 17–36.

Bird, N. 2011. Going Beyond Aid Effectiveness to Guide the Delivery of Climate Finance. ODI Background Note. London: Overseas Development Institute.

Bodansky, D. 2010. The Copenhagen Climate Change Conference: A Post-mortem. *American Journal of International Law* 104: 230–240.

Brooks, N., W.N. Adger, and P.M. Kelly. 2005. The Determinants of Vulnerability and Adaptive Capacity at the National Level and the Implications for Adaptation. *Global Environmental Change* 15: 151–163.

Brown, J., N. Bird, and L. Schalatek. 2010. Climate Finance Additionality: Emerging Definitions and Their Implications. Climate Finance Policy Brief No. 2. Heinrich Böll Stiftung. The Green Political Foundation. Available online at https://www.odi.org/sites/odi.org.uk/files/odi-assets/publications-opinion-files/6032.pdf.

Burton, I. 1994. Deconstructing Adaptation ... And Reconstructing. *Delta* 5 (1): 14–15.

Burton, I., S. Huq, B. Lim, O. Pilifosova, and E.L. Schipper. 2002. From Impacts Assessment to Adaptation Priorities: The Shaping of Adaptation Policy. *Climate Policy 2* (2–3): 145–159.

Carty, T., J. Kowalzig, and A. Peterson. 2016. Climate Finance Shadow Report 2016: Lifting the Lid on Progress Towards the $100 Billion Commitment. Oxfam International. Available online at http://www.oxfamnovib.nl/Redactie/Pdf/bp-climate-finance-shadow-report-031116-en.pdf.

Ciplet, D., J.T. Robert, and M. Khan. 2013. The Politics of International Climate Adaptation Funding: Justice and Divisions in the Greenhouse. *Global Environmental Politics* 13 (1): 49–68.

Clist, P. 2011. 25 Years of Aid Allocation Practice: Whither Selectivity? *World Development* 39 (10): 1724–1734.

DARA (2012). *Climate Vulnerability Monitor. A Guide to the Cold Calculus of a Hot Planet.* 2nd ed. Geneva: Fundación DARA Internacional.

Dasgupta, D., and Climate Change Finance Unit. 2015. Climate Change Finance, Analysis of a Recent OECD Report: Some Credible Facts Needed. Climate Change Finance Unit, Department of Economic Affairs, Ministry of Finance, Government of India. Available online at http://pibphoto.nic.in/documents/rlink/2015/nov/p2015112901.pdf.

Davies, R.B., and S. Klasen. 2013. Of Donor Coordination, Free-Riding, Darlings, and Orphans: The Dependence of Bilateral Aid Commitments on Other Bilateral Giving. CESifo Working Paper No. 4177.

Diamond, J., and C. Bruch. 2011. The International Architecture for Climate Change Adaptation Assistance. In *Climate Change Adaptation and International Development: Making Development Cooperation More Effective*, chap. 4.1, ed. R. Fujikura, and M. Kawanishi, 291–315. London/Washington, DC: Earthscan.

Dollar, D., and V. Levin. 2006. The Increasing Selectivity of Foreign Aid, 1984–2003. *World Development* 34 (12): 2034–2046.

Dreher, A., P. Nunnenkamp, and R. Thiele. 2008. Does Us Aid Buy un General Assembly Votes? A Disaggregated Analysis. *Public Choice 136* (1–2): 139–164.

Drury, A.C., R.S. Olson, and D.A. Van Belle. 2005. The Politics of Humanitarian Aid: U.S. Foreign Disaster Assistance, 1964–1995. *The Journal of Politics* 67 (2): 454–473.

Dudley, L., and C. Montmarquette. 1976. A Model of the Supply of Bilateral Foreign Aid. *The American Economic Review* 66 (1): 132–142.

Duus-Otterström, G. 2015. Allocating Climate Adaptation Finance: Examining Three Ethical Arguments for Recipient Control. *International Environmental Agreements* 16 (6): 655–670.

Eriksen, S.H., and P.M. Kelly. 2007. Developing Credible Vulnerability Indicators for Climate Adaptation Policy Assessment. *Mitigation and Adaptation Strategies for Global Change* 12: 495–524.

Fankhauser, S. 2010. The Costs of Adaptation. *WIRES Climate Change* 1: 23–30.
Fankhauser, S., and I. Burton. 2011. Spending Adaptation Money Wisely. *Climate Policy* 11: 1037–1049.
Figaj, M. 2010. Who Gets Environmental Aid? The Characteristics of Global Environmental Aid Distribution. *Environmental Economics and Policy Studies* 12 (3): 97–114.
Fuhr, L., L. Schalatek, and S. Ilse. 2016. 'We'll always have Paris'. Heinrich Böll Stiftung. The Green Political Foundation. Available online at http://www.boell.de/en/2016/12/01/well-always-have-paris.
Füssel, H.-M. 2009. Review and Quantitative Analysis of Indices of Climate Change Exposure, Adaptive Capacity, Sensitivity, and Impacts. Technical Report, Background Note to the World Development Report 2010.
Füssel, H.-M. 2010. How Inequitable is the Global Distribution of Responsibility, Capability, and Vulnerability to Climate Change: A Comprehensive Indicator-based Assessment. *Global Environmental Change* 20: 597–611.
Government of Sweden. 2014. Aid Policy Framework – The Direction of Swedish Aid. Government Communication 2013/14:131.
Grasso, M. 2010a. An Ethical Approach to Climate Adaptation Finance. *Global Environmental Change* 20: 74–81.
Grasso, M. 2010b. *Justice in Funding Adaptation under the International Climate Change Regime*. Dordrecht etc.: Springer.
Green Climate Fund. 2014. 3.3 Policies for Resource Allocation. 6th Board Meeting, Decision B.06/06, Bali, Indonesia.
Guillaumont, P., and C. Simonet. 2011. Designing an Index of Physical Vulnerability to Climate Change. Available online at http://www.ferdi.fr/sites/www.ferdi.fr/files/idi/B18.pdf. Ferdi Policy Brief.
Gulrajani, N. 2016. Bilateral Versus Multilateral Aid Channels: Strategic Choices for Donors. ODI Report. Available online at https://www.odi.org/sites/odi.org.uk/files/resource-documents/10492.pdf, Overseas Development Institute, London.
Gupta, J. 1997. *The Climate Change Convention and Developing Countries: From Conflict to Consensus?* Boston/London/Dordrecht: Kluwer Academic Publishers.
Gupta, J. 2010. A History of International Climate Change Policy. *WIRES Climate Change* 1: 636–653.
Gupta, J. 2014. *The History of Global Climate Governance*. Cambridge: Cambridge University Press.
Halimanjaya, A. 2013. Climate Finance Across Developing Countries: What Are the Major Determinants? DEV Working Paper 45. University of East Anglia, Norwich.
Halimanjaya, A. 2015. Climate Mitigation Finance Across Developing Countries: What Are the Major Determinants? *Climate Policy* 15 (2): 223–252.

Harmeling, S., and A.O. Kaloga. 2011. Understanding the Political Economy of the Adaptation Fund. *IDS Bulletin* 42 (3): 23–32.

Harris, P.G. 1999. Common But Differentiated Responsibility: The Kyoto Protocol and United States Policy. *New York University Environmental Law Journal* 7 (1): 27–48.

Hicks, R.L., B.C. Parks, J.T. Robert, and M.J. Tierney. 2008. *Greening Aid? Understanding the Environmental Impact of Development Assistance*. Oxford: Oxford University Press.

Hinkel, J. 2011. "Indicators of vulnerability and adaptive capacity": Towards a Clarification of the Science–Policy Interface. *Global Environmental Change* 21: 198–208.

Hoeffler, A., and V. Outram. 2011. Need, Merit, or Self-Interest—What Determines the Allocation of Aid? *Review of Development Economics* 15 (2): 237–250.

Horstmann, B. 2011. Operationalizing the Adaptation Fund: Challenges in Allocating Funds to the Vulnerable. *Climate Policy* 11 (4): 1086–1096.

Huhtala, A., S. Curto, and P. Ambrosi. 2010. Monitoring Climate Finance and ODA. World Bank Issues Brief 1.

Huq, S. 2016. Adaptation Finance: Climate Change's Forgotten Child. Climate Home. Available online at http://www.climatechangenews.com/2016/11/15/adaptation-finance-climate-changes-forgotten-child/.

Huq, S., and J.M. Ayers. 2008. Streamlining Adaptation to Climate Change into Development Projects at the National and Local Level. In *Financing Climate Change Policies in Developing Countries*, ed. European Parliament, Number PE 408.546- IP/A/CLIP/A/CLIM/ST/2008-13. European Parliament.

Kaly, U., C. Pratt, and J. Mitchell. 2004. The Environmental Vulnerability Index (EVI) 2004. Secretariat of the Pacific Community Applied Geoscience and Technology Division (SOPAC) Technical Report 384. Available online at http://www.sopac.org/sopac/evi/Files/EVI%202004%20Technical%20Report.pdf.

Kevlihan, R., K. DeRouen Jr., and G. Biglaiser. 2014. Is US Humanitarian Aid Based Primarily on Need or Self-Interest? *International Studies Quarterly* 58: 839–854.

Khan, M.R., and T.J. Roberts. 2013. Adaptation and International Climate Policy. *WIRES Climate Change* 4: 171–189.

Klasen, S., and R.B. Davies. 2011. Of Donor Coordination, Free-Riding, Darlings, and Orphans: The Dependence of Bilateral Aid Commitments on Other Bilateral Giving. Proceedings of the German Development Economics Conference, Berlin 2011, No. 47.

Klein, R.J. 2009. Identifying Countries that Are Particularly Vulnerable to the Adverse Effects of Climate Change: An Academic or a Political Challenge? *Carbon & Climate Law Review* 3 (3): 284–291.

Klein, R.J. 2010. Linking Adaptation and Development Finance: A Policy Dilemma Not Addressed in Copenhagen. *Climate and Development* 2 (3): 203–206.

Klein, R.J., and A. Möhner. 2011. The Political Dimension of Vulnerability: Implications for the Green Climate Fund. *IDS Bulletin* 42 (3): 15–22.

Lewis, T.L. 2003. Environmental Aid: Driven by Recipient Need or Donor Interest? *Social Science Quarterly* 84 (1): 144–161.

Lightfoot, S., G.A. Davies, and R. Johns. 2016. Needs and Interests: Understanding the British Public's Balancing of Aid Priorities. Paper presented at the Workshop 'The Domestic Dimensions of Development Cooperation', Institute for Development Policy and Management, University of Antwerp, Antwerp, Belgium, 24–25 October 2016.

McKinlay, R.D. 1978. The German Aid Relationship: A Test of the Recipient Need and the Donor Interest Models of the Distribution of German Bilateral Aid 1961–70. *European Journal of Political Research* 6: 235–257.

McKinlay, R.D., and R. Little. 1977. A Foreign Policy Model of U.S. Bilateral Aid Allocation. *World Politics* 30 (1): 58–86.

McKinlay, R.D., and R. Little. 1979. The U.S. Aid Relationship: A Test of the Recipient Need and Donor Interest Models. *Political Studies* 27 (2): 236–250.

Miller, D.C. 2014. Explaining Global Patterns of International Aid for Linked Biodiversity Conservation and Development. *World Development* 59: 341–359.

Moss, R.H., A.L. Brenkert, and E.L. Malone. 2001. Vulnerability to Climate Change: A Quantitative Approach. Technical Report, Pacific Northwest National Laboratory PNNL-SA-33642. Prepared for the U.S. Department of Energy.

Muccione, V., S.K. Allend, C. Huggel, and J. Birkmann. 2017. Differentiating Regions for Adaptation Financing: The Role of Global Vulnerability and Risk Distributions. *WIREs Climate Change* 8 (2): e447.

Nakhooda, S., T. Fransen, T. Kuramochi, A. Caravani, A. Prizzon, N. Shimizu, H. Tilley, A. Halimanjaya, and B. Welham. 2013. Mobilising International Climate Finance: Lessons from the Fast-Start Finance Period. Institute for Global Environmental Strategies, World Resources Institute, and Overseas Development Institute.

ND-GAIN (n.d.). ND-GAIN: Notre Dame Global Adaptation Index. Available online at http://index.gain.org/. Last accessed 15 March 2017.

OECD. 2011. Handbook on the OECD-DAC Climate Markers. Available online at www.oecd.org/dac/stats/48785310.pdf.

OECD. 2016a. DAC Glossary of Key Terms and Concepts. Available online at http://www.oecd.org/dac/dac-glossary.htm. Accessed 9 Nov 2016.

OECD. 2016b. Development aid rises again in 2015, spending on refugees doubles. Available online at http://www.oecd.org/dac/financing-sustainable-development/development-finance-standards/development-aid-rises-again-in-2015-spending-on-refugees-doubles.htm. Accessed 9 Nov 2016.

OECD. 2016c. Roadmap to US$100 Billion. Available online at http://dfat.gov.au/international-relations/themes/climate-change/Documents/climate-finance-roadmap-to-us100-billion.pdf.
Oxfam International. 2007. Adapting to Climate Change: What's Needed in Poor Countries, and Who Should Pay. Oxfam Briefing Paper 104. Available online at https://www.oxfam.org/sites/www.oxfam.org/files/adapting%20to%20climate%20change.pdf.
Parliament of the United Kingdom. 2015. International Development (Official Development Assistance Target) Act 2015. Chapter 12. Available online at http://www.legislation.gov.uk/ukpga/2015/12/pdfs/ukpga_20150012_en.pdf.
Parry, M., N.W. Arnell, M. Hulme, R. Nicholls, and M. Livermore. 1998. Adapting to the Inevitable. *Nature* 395, 741.
Persson, Å., and E. Remling. 2014. Equity and Efficiency in Adaptation Finance: Initial Experiences of the Adaptation Fund. *Climate Policy* 14 (4): 488–506.
Peterson Carvalho, A., and P. Terpstra, 2015. Tracking Adaptation Finance: An Approach for Civil Society Organizations to Improve Accountability for Climate Change Adaptation. Oxfam America Inc. and World Resources Institute. Available online at http://www.wri.org/publication/tracking-adaptation-finance.
Pielke, R.A. 1998. Rethinking the Role of Adaptation in Climate Policy. *Global Environmental Change* 8 (2): 159–170.
Remling, E., and Å. Persson. 2015. Who is Adaptation for? Vulnerability and Adaptation Benefits in Proposals Approved by the UNFCCC Adaptation Fund. *Climate and Development* 7 (1): 16–34.
Roberts, T.J., and R. Weikmans. 2015. The Unfinished Agenda of the Paris Climate Talks: Finance to the Global South. Brookings Series: Understanding COP21 and Beyond, Number 24. December 22. Available at: http://www.brookings.edu/blogs/planetpolicy/posts/2015/12/22-cop21-unfinished-agenda-finance-global-south-roberts-weikmans.
Roberts, J.T., and R. Weikmans. 2017. Postface: Fragmentation, Failing Trust and Enduring Tensions over What Counts as Climate Finance. *International Environmental Agreements: Politics, Law and Economics* 17 (1): 129–137.
Roberts, T.J., R. Weikmans, and C. Jones. 2017. Allocating Climate Adaptation Finance: Shifting Positions in the Scholarly and Policy Realms.
Robertsen, J., N. Francken, and N. Molenaers. 2015. Determinants of The Flow of Bilateral Adaptation-Related Climate Change Financing to Sub-Saharan African Countries. LICOS Discussion Paper 373/2015. Catholic University Leuven.
Robinson, S.-A., and M. Dornan. 2017. International Financing for Climate Change Adaptation in Small Island Developing States. *Regional Environmental Change*. https://doi.org/10.1007/s10113-016-1085-1.

Schipper, E.L.F. 2006. Conceptual History of Adaptation in the UNFCCC Process. *RECIEL: Review of European Community & International Environmental Law* 15 (1): 82–92.

Sharma, A. 2017. Precaution and Post-caution in the Paris Agreement: Adaptation, Loss and Damage and Finance. *Climate Policy* 17 (1): 33–47.

Smit, B., and J. Wandel. 2006. Adaptation, Adaptive Capacity and Vulnerability. *Global Environmental Change* 16 (3): 282–292.

Smith, J.B., T. Dickinson, J.D.B. Donahue, I. Burton, E. Haites, R.J. Klein, and A. Patwardhan. 2011. Development and Climate Change Adaptation Funding: Coordination and Integration. *Climate Policy* 11: 987–1000.

Stadelmann, M., J.T. Roberts, and A. Michaelowa. 2010. Keeping a Big Promise: Options for Baselines to Assess "New and Additional" Climate Finance. Centre for Comparative and International Studies (ETH Zurich and University of Zurich) Working Paper No. 66.

Stadelmann, M., Å. Persson, I. Ratajczak-Juszko, and A. Michaelowa. 2014. Equity and Cost-Effectiveness of Multilateral Adaptation Finance: Are They Friends or Foes? *International Environmental Agreements: Politics, Law and Economics* 14: 101–120.

Stern, N.H. 2009. *Blueprint for a Safer Planet. How to Manage Climate Change and Create a New Era of Progress and Prosperity*. London: Random House.

Stone, C.D. 2004. Common But Differentiated Responsibilities in International Law. *American Journal of International Law* 98: 276–301.

Turner, B.L., R.E. Kasperson, P.A. Matsone, J.J. McCarthy, R.W. Corell, L. Christensen, N. Eckley, J.X. Kasperson, A. Luers, M.L. Martello, C. Polsky, A. Pulsiphera, and A. Schiller. 2003. A Framework for Vulnerability Analysis in Sustainability Science. *Proceedings of the National Academy of Sciences* 100 (14): 8074–8079.

UNDP. 2007. Human Development Report 2007/2008. Fighting Climate Change: Human Solidarity in a Divided World. United Nations Development Programme, New York. Available online at http://hdr.undp.org/sites/default/files/reports/268/hdr_20072008_en_complete.pdf.

UNFCCC. 1992. United Nations Framework Convention on Climate Change. Document FCCC/INFORMAL/84.

UNFCCC. 1995. Report of the Conference of the Parties on its first session, held at Berlin from 28 March to 7 April 1995. Addendum Part Two: Action taken by the Conference of the Parties at its first session. Document FCCC/CP/1995/7/Add.1.

UNFCCC. 1997. Kyoto Protocol to the United Nations Framework Convention on Climate Change. Decision 1/CP.3. Document FCCC/CP/1997/7/Add.1.

UNFCCC. 2001. Report of the Conference of the Parties on its Seventh Session, held at Marrakesh from 29 October to 10 November 2001. Addendum. Part Two: Action Taken by the Conference of the Parties. Document FCCC/CP/2001/13/Add.1.

UNFCCC. 2003. Report of the Conference of the Parties on Its Ninth Session, held at Milan from 1 to 12 December 2003. Addendum. Part Two: Action Taken by the Conference of the Parties at Its Ninth Session. Document FCCC/CP/2003/6/Add.1.

UNFCCC. 2004. Report of the Conference of the Parties on Its Tenth Session, held at Buenos Aires from 6 to 18 December 2004 Addendum Part Two: Action Taken by the Conference of the Parties at Its Tenth Session. Document FCCC/CP/2004/10/Add.1.

UNFCCC. 2007. Report of the Conference of the Parties on Its Thirteenth Session, held in Bali from 3 to 15 December 2007 Addendum Part Two: Action Taken by the Conference of the Parties at its thirteenth session. Document FCCC/CP/2007/6/Add.1.

UNFCCC. 2008. Report of the Conference of the Parties Serving as the Meeting of the Parties to the Kyoto Protocol on its Fourth Session, held in Poznan from 1 to 12 December 2008. Addendum. Decision 1/CMP.4: Adaptation Fund. FCCC/KP/CMP/2008/11/Add.2.

UNFCCC. 2009. Report of the Conference of the Parties on Its Fifteenth Session, held in Copenhagen from 7 to 19 December 2009. Addendum. Part Two: Action Taken by the Conference of the Parties at Its Fifteenth Session. Document FCCC/CP/2009/11/Add.1.

UNFCCC. 2010. The Cancún Agreements: Outcome of the Work of the Ad Hoc Working Group on Long-term Cooperative Action Under the Convention. Decision 1/CP.16. Document FCCC/CP/2010/7/Add.1.

UNFCCC. 2011. Report of the Conference of the Parties on its Seventeenth Session, held in Durban from 28 November to 11 December 2011. Addendum. Part Two: Action Taken by the Conference of the Parties at Its Seventeenth Session. Document FCCC/CP/2015/10/Add.1.

UNFCCC. 2015. Adoption of the Paris Agreement. Decision 1/CP.21. Document FCCC/CP/2015/10/Add.1.

UNFCCC. 2016. Report of the Conference of the Parties on Its Twenty-Second Session, held in Marrakech from 7 to 18 November 2016. Addendum. Part Two: Action Taken by the Conference of the Parties at Its Twenty-Second Session. FCCC/CP/2016/10/Add.1.

United Nations. 2002. Monterrey Consensus of the International Conference on Financing for Development. The Final Text of Agreements and Commitments Adopted at the International Conference on Financing for Development, Monterrey, Mexico, 18–22 March 2002. Available online at http://www.un.org/esa/ffd/wp-content/uploads/2014/09/MonterreyConsensus.pdf.

Weikmans, R. 2016a. Dimensions éthiques de l'allocation du financement international de l'adaptation au changement climatique. *VertigO - la revue électronique en sciences de l'environnement 16* (2).

Weikmans, R. 2016b. Le rôle de la coopération au développement dans le financement international de l'adaptation au changement climatique. In *La nouvelle géographie du développement. Coopérer dans un monde en mutation*, ed. A. Zacharie, 175–185. Brussels: Éditions Le Bord de l'Eau.

Weiler, F., C. Betzold, and M. Dornan. 2017. Vulnerability or Donor Interests? The Allocation of Aid for Climate Change Adaptation. Paper presented at the Comparative Politics Seminars, 5 April 2017, University of Kent, Canterbury, UK

Wheeler, D. 2011. Quantifying Vulnerability to Climate Change: Implications for Adaptation Assistance. Centre for Global Development Working Paper 240.

World Bank. 2006. Managing Climate Risk: Integrating Adaptation into World Bank Group Operations. Washington, DC.

Younas, J. 2008. Motivations for Bilateral Aid Allocation: Altruism or Trade Benefits. *European Journal of Political Economy* 24: 661–674.

Zanger, S.C. 2000. Good Governance and European Aid: The Impact of Political Conditionality. *European Union Politics* 1 (3): 293–317.

CHAPTER 3

Setting the Stage: A Mixed Methods Research Design

Now that we have provided the backdrop of our study—the international climate change negotiations—and embedded our research in the broader aid allocation literature, we can turn to the core of our study: the empirical analysis. We want to understand how donors distribute their adaptation-relevant development assistance: who receives support for climate change adaptation, and why? We address this question using both quantitative and qualitative research methods. On the one hand, we need quantitative methods to detect patterns in adaptation aid giving across all OECD DAC donors and recipients—we examine adaptation aid flows from 28 donors to 141 recipients. On the other hand, we also need qualitative methods to understand the decision-making processes that drive the patterns detected by our quantitative analysis. The mix of methods also helps to compensate the weaknesses of the different methods, such as data limitations in quantitative datasets. We hence combine statistical analysis of OECD aid data with qualitative case studies of three large climate donors: Germany, Sweden, and the UK. We selected these three donors using a very different systems design, as Sect. 3.1 explains. Although Germany, Sweden, and the UK are all rather generous supporters of climate change adaptation, they differ in many other regards, such as their size, economic structure, or colonial history.

We use regression analysis to examine systematically whether vulnerability, good governance, or donors' foreign policy interests drive adaptation aid-giving, in these three countries as well as across all OECD donors. We

C. Betzold and F. Weiler, *Development Aid and Adaptation to Climate Change in Developing Countries*,
https://doi.org/10.1007/978-3-319-64510-0_3

operationalise the different potential drivers of adaptation aid allocation through different indicators. As we describe in Sect. 3.2 in greater detail, we use for instance three indicators of physical vulnerability: the exposure sub-index of the Notre Dame Global Adaptation Index, the Environmental Vulnerability Index, and the Climate Risk Index.

Finally, Sect. 3.3 turns to our qualitative methods. The qualitative case studies shed light on the decision-making process of aid allocation and position adaptation aid in the wider official development assistance landscape of the three countries. To this end, we conducted semi-structured interviews with aid practitioners and stakeholders in the three case study countries and analysed key policy documents.

3.1 Our Overall Research Design: Mixed Methods

Both qualitative and quantitative methods are valid ways to study political phenomena such as adaptation finance. These two different research strategies should not be understood as rigid opposites to each other, but rather as two ends of a continuous spectrum (Newman and Benz 1998). In the middle of the continuum are mixed methods approaches, which incorporate elements of both quantitative and qualitative research design. Mixed methods approaches assume 'that the combination of qualitative and quantitative approaches provides a more complete understanding of a research problem than either approach alone' (Creswell 2013, 3).

To obtain a picture of adaptation aid allocation that is as complete as possible, we opted for a mixed methods research design in this book. Specifically, we combine a large-n quantitative study across all OECD donor countries on the one hand with a more detailed look at how three carefully chosen donor countries distribute their adaptation aid: Germany, Sweden, and the UK. For these three donors, we first run separate statistical models and then further investigate adaptation allocation through in-depth case studies using semi-structured interviews and key policy documents. Thus, our empirical analysis proceeds in three steps. We start out with a very broad look at the overall quantitative picture of adaptation aid flows across all donors using descriptive statistics in Chap. 4 and regression analysis in Chap. 5. In a second step, we zoom in to our three case study countries. We start with a statistical analysis of adaptation aid flows in the three selected countries (also in Chap. 5) and then use qualitative methods to examine aid allocation decisions in more detail and to add context to the findings of our statistical analyses in Chap. 6. The present chapter describes the quantitative and qualitative design used in this book and the respective data sources.

As mentioned above, we selected our three country cases using a very different systems design. The goal in such a design is to select cases that vary in as many characteristics as possible: 'the strategy is to choose units of research which are as different as possible with regard to extraneous variables. The basic logic is that differences cannot explain similarities' (Anckar 2008, 390).

These similarities between the three case study countries concern the level of development aid they provide. Germany, Sweden, and the UK are comparatively large international aid donors, in terms of both development aid in general and adaptation and climate aid in particular.[1] In 2015, the UK was the second largest aid donor in absolute terms (after the USA), with about $18.7 billion in total development aid. Germany followed closely in third place with about $17.8 billion in total development aid. Sweden is much smaller in population, but spends the highest percentage in terms of GNI of all countries on development cooperation: about 1%. Sweden was the sixth largest donor in absolute terms with $7.1 billion given to development cooperation in 2015 (for more details see OECD 2016a). Like Sweden, the UK has reached the United Nations ODA target of spending 0.7% of GNI on development cooperation, which Germany has not (see OECD 2016b; Chap. 6). Furthermore, all three countries have a historical commitment to providing development aid, backed by the countries' parliaments and—to some degree—public opinion (see e.g. Olsen 2001). As Chaps. 4 and 6 show in greater detail, Germany, Sweden, and the UK are also among the most important providers of aid for adaptation, yet as we will see, the distribution of adaptation aid to recipient countries varies between them in terms of scope and reach.

These differences may relate to differences across the three countries: Germany is a federal and comparatively conservative country, the UK is a strongly centralised country with a more libertarian outlook, while Sweden is a smaller yet traditionally social-democratic country. Germany, as a large exporter, additionally has strong economic ties across the world. The UK has strong political ties with many developing countries, thanks to its many former colonies and the Commonwealth. Sweden, finally, neither is a large exporter nor a former colonial power. Additionally, and of particular importance for our study, the three selected countries differ in how aid allocation is organised. Germany operates a relatively 'centralised' system with much of its development aid being distributed through—and projects implemented by—government funded agencies, notably the Gesellschaft für Internationale Zusammenarbeit (Society for International Cooperation, GIZ) for technical cooperation and the KfW for financial cooperation. While Sweden's aid agency, the Swedish International Development Cooperation Agency (Sida), also works on the ground, the country also strongly relies on external partners, notably non-governmental organisations (NGOs), some of which have preferential status. The UK does not have its own aid agency equivalent to the GIZ, KfW, or Sida, but works through partner organisations. These differences in socio-economic structure as well as in development aid structure help us understand potential differences in adaptation aid allocation and also add to the representativeness and generalisability of our results.

In the next two sections we describe both the quantitative and the qualitative research design in more detail. We start with the quantitative design and describe all the data sources we used to compile the dataset utilised in our regression analysis. Finally, we describe the qualitative research design and detail our data sources—semi-structured interviews and policy documents—as well as our method of coding and analysing the qualitative data.

3.2 Quantitative Design

We start our empirical analysis by examining the distribution of adaptation aid in the period 2010 through 2015, first across 28 OECD DAC donors and then separately for the three selected case study countries. After outlining the basic set-up of our dataset, we present the OECD CRS aid data used to construct the dependent variable for our analysis—adaptation aid—and discuss potential problems this data presents for our endeavour. We then describe the operationalisation of the various independent variables used in our study and finally explain our modelling strategy.

To study bilateral adaptation aid allocation we have to consider on the one hand the donors who provide adaptation aid, and on the other hand the recipients who receive this aid. In addition, aid allocation decisions can vary on a yearly basis, which is why we include time as a third dimension in the analysis. Thus we developed a fully dyadic dataset containing yearly adaptation aid flows from all OECD donors to all eligible recipients of adaptation aid, that is, countries on the OECD list of ODA recipients (see Chap. 1).[2] In other words, the unit of observation of our quantitative analysis is the donor–recipient–year triplet. Here, we describe in detail how we build this dataset, which data sources we use, and which variables we include to test our hypotheses.

Our analysis covers the period 2010 through 2015, for which we have OECD data on adaptation aid (see below). The quantitative analysis includes 28 donor countries that started to provide bilateral adaptation aid before 2013. For two of these countries (the Czech Republic and Iceland), data is only available for 5 years (from 2011 through 2015), while we obtained data for the entire time horizon of the study from 2010 through 2015 for all other 26 adaptation aid donors. These 26 donors are, in alphabetical order: Australia, Austria, Belgium, Canada, Denmark, Finland, France, Germany, Greece, Ireland, Italy, Japan, Kuwait, Luxembourg, the Netherlands, New Zealand, Norway, Portugal, Slovenia, South Korea, Spain, Sweden, Switzerland, the United Arab Emirates, the UK, and the USA. On the recipient side, the dataset includes 141 recipient countries.

As described in detail in Chap. 2, we assess three key factors that potentially influence the distribution of adaptation aid: recipient need, recipient merit, and donor interests. From a data-centric point of view, recipient need and recipient merit are characteristics of the recipient country. The value of a climate change vulnerability indicator or a good governance indicator for, say, Bangladesh is uniform across donors. Accordingly, we would expect *all* donors to give more adaptation aid to Bangladesh if they provided their adaptation aid based on recipient need and recipient merit only. This would not require a dyadic dataset; a recipient-based pooled dataset would suffice.

When measuring donor interests, on the other hand, we clearly need to consider both the donor and the recipient side. For instance, if we want to test whether donors provide adaptation aid to those countries with which they trade a lot, bilateral trade flows are relevant. This makes the development of a fully dyadic dataset necessary. To illustrate: trade flows between Bangladesh and the various donors differ. Accordingly, we would not expect *all* donors to give more adaptation aid to Bangladesh if they provided their aid based on donor interests. Let us assume that Bangladesh trades a lot with the UK, but very little with Sweden. According to our donor interest hypothesis, we would hence expect that Bangladesh received more adaptation aid from the UK than from Sweden. We can only test whether this is the case with a dyadic dataset.

3.2.1 Our Dependent Variable: Adaptation Aid

The core of our dyadic dataset is adaptation aid—we want to explain the variation in the level of adaptation aid that countries receive. Data on adaptation aid comes from the OECD CRS, which collects aid data from all OECD DAC donor countries. In 1998, the OECD DAC introduced so-called Rio markers to monitor the mainstreaming of environmental objectives in development cooperation (see e.g. Brown et al. 2010; OECD 2011). Originally, there were three markers related to the three UN Conventions on climate change, biodiversity, and desertification signed at the 1992 Earth Summit. In 2009, the climate change marker was split into a marker for climate change mitigation and a marker for climate change adaptation, or adaptation marker (OECD 2011). Rio markers are at the project level: donors report for each and every aid project whether it contributes to reducing greenhouse gas emissions, to dealing with the consequences of climate change, to conserving biodiversity, and/or to fighting desertification.

The adaptation marker defines an aid activity as relevant for adaptation if it 'intends to reduce the vulnerability of human or natural systems to the impacts of climate change and climate-related risks, by maintaining or increasing adaptive capacity and resilience' (OECD 2011, 4). As with all Rio markers, the adaptation marker has four possible values: a project may not be screened for its adaptation relevance or is screened but deemed unrelated to climate change adaptation. In both these cases, we consider the project as irrelevant for adaptation—that is, we consider the project as not intending to reduce vulnerability to climate risks—and do not take it into account. Projects that are relevant for adaptation may have adaptation as their *principal* objective on the one hand, or as a *significant* objective on the other. Projects with adaptation as their principal objective would not have taken place if it was not for the specific purpose of adaptation to climate change. In other words, a principal adaptation project is first and foremost about reducing vulnerability and increasing adaptive capacity and resilience. In contrast, projects with adaptation as a significant objective are largely about something else (for instance biodiversity conservation), but still make an important contribution to climate change adaptation (OECD 2011, 5).

Note that the Rio markers are not mutually exclusive; a project may thus at the same time have climate change adaptation and mitigation as its objective, for example. To illustrate, we can imagine a fictitious development project by the USA which sets up solar panels in communities in Togo. Let us say that the project at the same time includes an information and awareness campaign about climate change and dealing with climate risks. In this case, the USA would probably indicate that climate change mitigation was the project's principal objective. Since the project also includes an adaptation dimension, the USA would hence also assign it a significant adaptation objective in the OECD CRS.

As this example shows, aid projects in the OECD CRS often have multiple objectives, and adaptation may be just a small part of the overall project. The fictitious American project setting up solar panels in Togo and running an information and awareness campaign illustrates this: of the overall $500,000 committed to the project, most resources will likely go into purchasing solar panels, setting them up, and training locals to maintain them. Just a small part of project expenses—maybe $150,000—is used to pay for workshops, radio spots, and other elements of the information and awareness campaign. Since adaptation-specific investments may be only part of a broader project, as in this fictitious example, it is very difficult to identify, track, and quantify the exact amount of funding targeting climate change adaptation, as the OECD itself recognises (OECD 2016c, 11).

That Rio markers are not mutually exclusive and projects are double-counted towards multiple objectives is just one problem of the OECD aid data. Another problem concerns self-reporting: the dataset relies entirely on donors' own classifications of projects, which is inherently problematic. Self-reporting leads to inconsistent application of the adaptation marker and over-reporting. The OECD definition of adaptation is rather vague: an adaptation activity reduces vulnerability 'by maintaining or increasing adaptive capacity and resilience' (OECD 2011, 4). Donors and project managers within a single donor country likely understand and interpret this definition in different ways. As a result, application of the adaptation marker may be inconsistent. In general, donors tend to interpret adaptation in a rather broad sense, which leads to over-reporting. Several studies found that donors tend to overstate how much of their aid is relevant for climate change, including adaptation, and that this was particularly the case for projects with significant adaptation objectives (Adaptation Watch 2015; Junghans and Harmeling 2012; Michaelowa and Michaelowa 2011). Donors may feel pressured to commit large amounts of aid to climate change to fulfil their international pledges and thus interpret this definition rather broadly (Donner et al. 2016; Pickering et al. 2015; see also Chaps. 2 and 6). Yet even the opposite—under-reporting—is conceivable because 'donors may intentionally not label aid that contributes to adaptation as "climate" because of internal politics surrounding climate change' (Donner et al. 2016, 7). In other words, if the domestic political climate is unfavourable towards climate change policies, donors may decide not to label aid flows as adaptation aid, although they contribute to dealing with climate change impacts. In practice, over-reporting seems to be the bigger problem (Adaptation Watch 2015; Donner et al. 2016; Junghans and Harmeling 2012; Michaelowa and Michaelowa 2011). The numbers in the OECD CRS hence represent an optimistic upper bound of actual adaptation aid flows and should be interpreted with caution. Yet, there is little reason to expect systematic patterns in over-reporting: all donors probably overestimate the adaptation relevance of their aid. Over-reporting per se should therefore not affect our analysis: although the overall amount of adaptation aid is very likely smaller than what the OECD CRS reports, its distribution should not be systematically different because of over-reporting.

Clearly, the OECD CRS and its Rio markers are far from perfect. Many of the concerns of aid data in general also apply to adaptation aid (see Roodman 2007). Nonetheless, the OECD CRS data represent to date the most comprehensive and comparable data on development aid flows and their (perceived) relevance for adaptation objectives. While definitions and guidelines for applying the adaptation marker are ambiguous, at least the OECD has agreed definitions and methodologies, which the UNFCCC Biennial Reports for example lack. Further, most donors draw on their OECD data when reporting climate finance to the UNFCCC (Francke Lund et al. 2015, 36f). Nonetheless, the weaknesses of the OECD data should be kept in mind.

Our descriptive analysis in the next chapter makes use of the adaptation marker to explore development aid and adaptation aid flows with regard to time trends, differences across donors and recipients, or aid channels. Our statistical analysis in Chap. 5 then turns to a more systematic test of how adaptation aid is distributed. In the remainder of this section, we describe the variables used in the statistical analysis.

Our dependent variable is the level of adaptation aid a recipient country receives per year from each individual donor, including (but not limited to) our three selected donor countries, Germany, Sweden, and the UK. In mathematical terms, the dependent variable is A_{ijt}: the amount of adaptation aid A that recipient country i receives from donor j in year t.

To construct this variable, we first have to make a decision on how to treat projects where adaptation is the principal objective compared to projects where adaptation is only a significant objective. Although significant adaptation aid projects contribute to adaptation, they also would have happened in the absence of adaptation objectives. We therefore decided to use two different specifications of the dependent variable: one that includes principal adaptation aid only, and one that includes principal adaptation aid as well as significant aid. In this second specification, however, we discount significant adaptation aid projects at 50%, while using the full value of principal adaptation aid projects. Let us again take the fictitious American development project setting up solar panels in Togo and running an information and awareness campaign to illustrate. Suppose that the USA committed $500,000 to this project, for which adaptation is a significant objective. We would exclude this project from our first dependent variable (principal adaptation aid only) but include it in our second dependent variable—but discount it at 50%, that is, we would only include the second project at a value of $250,000. Many donors report their climate finance from ODA budgets applying such a discount factor, though the discount factors used vary (Adaptation Watch 2015, 32). By discounting significant adaptation aid at 50% we acknowledge the limitations of the OECD data, in particular issues of over-reporting and multiple objectives.

How then do we construct our two dependent variables? In a first step, we aggregate the OECD CRS project-level data for each recipient and each donor for each year, separately for principal and for significant adaptation aid. Thus, the dataset is fully dyadic. In other words, each donor–recipient pair has two entries for every year of adaptation aid flows: one entry that sums up all *principal* adaptation aid that a recipient *i* received from donor *j*, and one entry that sums up all *significant* adaptation aid that recipient *i* received from donor *j*. If a donor did not provide any adaptation aid to a specific recipient in a given year, the entry for this donor–recipient country pair is zero. In a second step, we calculate per capita adaptation aid allocated in a given year. In other words, we compute how much adaptation aid, on average, each citizen of a recipient country received from a particular donor, again separately for principal adaptation aid as well as for significant adaptation aid. We do this to make adaptation aid flows more comparable, since adaptation aid flows of, say, $10 million might seem almost insignificant in a country like China, while a similar project in a small island state can have a huge impact (see also Chap. 4). All aid data are available from the OECD Statistics webpage (see OECD 2016b).

To illustrate and to help readers better understand the two dependent variables just described, we can again turn to the fictitious American solar panel project. Let us assume that in 2012, the USA committed resources to only two adaptation-related development projects in Togo: the solar panel project we already know of, where adaptation is a significant objective, as well as a project that replants mangroves to reduce the risk of flooding from sea-level rise and storms. Let us say that this second project is worth $1 million and has adaptation as its principal objective. The entry for the triplet USA–Togo–2012 for our first dependent variable that considers only principal adaptation aid would thus be $1 million. The population of Togo in 2012 was about 6.7 million, so each Togolese received about 15 cents for principal adaptation from the USA in 2012. Our second dependent variable also includes the solar panel project, at 50% of its value, that is, at $250,000. In this example, the USA thus gave $1.25 million for principal and significant adaptation in Togo in 2012, or 19 cents for each Togolese. The two values of our dependent variables for the triplet USA–Togo–2012 would thus be 15 cents and 19 cents, respectively (assuming that the USA only funded two projects in Togo during that year).

To summarise the discussion, we have the following two dependent variables on how much adaptation aid recipient countries receive:

- Principal adaptation aid per capita that recipient *i* receives from donor *j* in year *t*;
- Principal and significant adaptation aid (the latter discounted at 50%) per capita that recipient *i* receives from donor *j* in year *t*.

3.2.2 Physical Vulnerability

Having specified our dependent variables, we now turn to our independent variables measuring recipient need, recipient merit, and donor interests. Recipient need is our first key independent variable. Recipient need, in an adaptation context, means vulnerability to climate change impacts. In line with our first hypothesis H1a, we expect that more vulnerable countries receive more adaptation aid. But which countries are more vulnerable?

Vulnerability is an inherently complex concept with no single definition, yet most scholars agree that it has two dimensions: (physical) exposure and sensitivity to natural hazard (H1a) on the one hand, and (socio-economic) adaptive capacity (H1b) on the other (e.g. Smit and Wandel 2006; see also Chap. 1). Vulnerability is hard—some would even say impossible—to measure (e.g. Engle 2011; Hinkel 2011). Nevertheless, there is a range of indices that seek to quantify vulnerability at the national level. Quantitative vulnerability indices are contested. The key problem is that vulnerability is multi-dimensional (see Chap. 2). Brooks et al. (2005) examine which elements might fall under a comprehensive index of vulnerability to climate change. The authors identify an initial shortlist of 46 variables that could potentially be used to measure climate change vulnerability, representing economic well-being and inequality, health and nutritional status, education, physical infrastructure, governance, geographic and demographic factors, agriculture, ecosystems, and technological capacity (Brooks et al. 2005, 153f). This list is almost too long to be useful, and includes many variables that are strongly correlated with each other (for instance a long list of proposed health, education, and economic variables), and that could also be used to capture donor interests and recipient merit (specifically the various governance indicators proposed by the authors).

Since we would like to test the role of physical vulnerability and adaptive capacity separately (H1a and H1b, see Chap. 2), we use indicators that clearly separate physical exposure and sensitivity to climate change from adaptive capacity. We start by focusing on the physical dimension of vulnerability to operationalise H1a. Yet even this is not straightforward, and a host of different indices of climate change vulnerability have been proposed, some focusing more on (short-term) weather events and their direct impacts on the livelihoods of people, others capturing the potential long-term consequences of climate change such as sea level rise or changing rainfall patterns. To circumvent this problem and to capture different dimensions of physical vulnerability, we include three different indicators of physical vulnerability in our analysis: the exposure sub-index of the Notre Dame Global Adaptation Index, the Environmental Vulnerability Index, and the Climate Risk Index. We discuss these three indicators in turn.

Notre Dame Global Adaptation Index

Our first indicator of physical vulnerability is the exposure sub-index of the University of Notre Dame's Global Adaptation Index (ND-GAIN). The ND-GAIN 'summarizes a country's vulnerability to climate change and other global challenges in combination with its readiness to improve resilience' (ND-GAIN n.d.). This sentence, taken from ND-GAIN's webpage, already demonstrates that the overall index is problematic for our purpose, as it blends together not only vulnerability and 'readiness', but also physical exposure and sensitivity to climate risks and adaptive capacity. More specifically, the ND-GAIN has two sub-indices: 'vulnerability' and 'readiness'. 'The vulnerability portion of ND-GAIN includes both climate risk ("exposure" and "sensitivity") and adaptive capacity', while 'readiness measures the ability of a country's private and public sectors to absorb investment resources and successfully apply them to reduce climate change vulnerability' (ND-GAIN 2013, 5; 7). To measure physical vulnerability to climate change, the two components 'exposure' and 'sensitivity' are of interest. Fortunately, ND-GAIN allows users of the index to download separately all sub-indices and variables used to construct the overall index.

We use the exposure sub-index only in our quantitative models, as this variable—in our view—best captures physical vulnerability. To illustrate and to make our choice more transparent, let us take a closer look at the ND-GAIN exposure and sensitivity sub-indices. The overall ND-GAIN aggregates a wide range of variables to measure vulnerability. The variables cover different sectors for both the exposure and the sensitivity sub-indices (plus adaptive capacity, see below), namely: water, food, health, human habitat, ecosystem services, and infrastructure. While the variables for the exposure sub-index all focus on the concept of physical exposure (see ND-GAIN 2013 for a list of all variables), the sensitivity sub-index contains problematic variables. For instance, the sector 'health' is operationalised by the variables 'health workers per capita' and 'health expenditures'; the sector 'infrastructure' (transport) is operationalised by the variable 'paved roads'. In our view, these variables could equally well capture adaptive capacity, or to some extent recipient merit, as they are measures of economic and organisational development in a country. Therefore, we rely entirely on the exposure sub-index, which in our view comes closest to the concept we are after.

In general, the ND-GAIN exposure sub-index captures the long-term consequences of climate change, as the underlying variables capture general trends such as rainfall patterns, impacts of future climate change on health, or areas in risk of flooding. Singular events such as a flood or a storm in a given year are not captured in this way, and do not play a role for the construction of the sub-index. This can also be seen by the relatively stable values of the exposure (as well as the sensitivity) sub-index over time. In short, the ND-GAIN exposure sub-index is a strong measure of physical vulnerability to long-term climate change.

The ND-GAIN has been available since 1995 and reports values for all sub-indices and variables for every year. The exposure sub-index has been computed to range theoretically from 0 to 1, with higher values representing higher levels of physical exposure. In practice, the range of the variable is relatively small, with a minimum of 0.36 for Moldova and a maximum of 0.74 for the Maldives.

Environmental Vulnerability Index

Our second measure of physical vulnerability is the Environmental Vulnerability Index (EVI). The South Pacific Applied Geoscience Commission (SOPAC) created the EVI in 2004, which includes a climate change sub-index. The EVI climate change sub-index is one of the few indicators of vulnerability to climate change which considers that part of vulnerability that neither depends on present nor on future policy:

> The EVI has been designed to reflect the extent to which the *natural environment* of a country is prone to damage and degradation. It does *not* address the vulnerability of the social, cultural or economic systems, nor the environment that has become dominated by those same human systems (such as cities and farms) because these are included in the economic and social vulnerability indices which are needed separately to identify trade-offs. Therefore, the natural environment includes those biophysical systems that can be sustained without direct and/or continuing human support (Kaly et al. 2004, 6; emphasis added).

The focus on the natural environment makes the EVI an ideal candidate for our purpose of capturing physical vulnerability to climate change separately from adaptive capacity.

Three main components together form the EVI: first, the likelihood that certain environmental hazards come into play for a country (hazards); second, the degree to which the natural environment is able to cope with these hazards (resistance); and third, the already sustained damages due to changes in the natural environment in the past (damage) (Kaly et al. 2004; SOPAC 2004). The full EVI is composed of 50 indicators measuring environmental vulnerability, and not only considers climate change impacts. In total, these 50 indicators build seven thematic sub-indices of the EVI for exposure to natural disasters, biodiversity, desertification, water, agriculture and fisheries, human environmental health, as well as climate change. This climate change sub-index, for which we will use the abbreviation EVI as a short-hand henceforth, is made up of 13 indicators: 6 pertaining to hazards (high winds, dry periods, wet periods, hot periods, sea temperatures, renewable water); 4 pertaining to resistance (land area, country dispersion, relief, lowlands); and 3 pertaining to damage (natural vegetation cover remaining, human population density, coastal settlements) (see Kaly et al. 2004 for detailed descriptions).

The individual components as well as the thematic sub-indices of the EVI are scaled to range from 1 to 7, with higher values indicating higher vulnerability. For the countries in our dataset, the climate change sub-index ranges from 1.67 for Botswana to 5.13 for St Kitts and Nevis.

The EVI is a long-term index of vulnerability and thus was computed only once (instead of providing values for different years). Other than for example the ND-GAIN index, which is based on yearly updated data and published annually, the EVI assigns only a single vulnerability value to each country. As such, this index is similar to the Structural Vulnerability to Climate Change Index (SVCCI) published by the Fondation pour les Etudes et Recherches sur le Développement International (Guillaumont and Simonet 2011). We decided to use the EVI instead of the newer SVCCI for two reasons: first, the EVI is more widely used and thus better known (and tested) in the literature; second, the EVI has broader coverage. It is available for 137 of the 141 recipient countries included in our dyadic dataset, while the SVCCI is only reported for 118 of these countries.

Climate Risk Index

Our final measure of physical vulnerability is the Global Climate Risk Index (CRI). The CRI, published by the German NGO Germanwatch, captures a country's annual exposure to weather related loss events such as floods or storms.[3] The index is based on data from Munich Re—according to the authors, 'one of the most reliable data sets available on the impacts of extreme weather events and associated socio-economic data' (Kreft et al. 2016, 3). The CRI, in contrast to the ND-GAIN exposure sub-index and EVI described earlier, reflects vulnerability to short-term weather related events or climate variability, rather than to the long-term effects of climate change. The annual reports accompanying the index also clearly state that the index does *not* consider important long-term effects of climate change, such as sea-level rise, ocean acidification, or glacier melting (e.g. Kreft and Eckstein 2014; Kreft et al. 2016).

To compile the index, Germanwatch collects data on large-scale weather-related events, and records for every event the death toll and economic losses in US$ at purchasing power parity (PPP). The CRI for a given country in a given year considers all events that occurred over the prior 20 years. In a second step, Germanwatch computes annual averages for the death toll, the death toll per 100,000 inhabitants, total economic losses in million US$ (PPP), and economic losses per unit GDP in percent (Anemüller et al. 2006). For instance, the index for 2014 is based on these four loss indicators from 1995 to 2014. Thus, consecutive years have a 19-year overlap in their data basis, and the values for individual countries do not vary greatly on an annual basis.

The CRI index records values for countries such that values closer to zero indicate *higher* vulnerability levels than larger values. The country most affected by climate change over the time horizon of the study—according to the index—was the Philippines in 2013 with a score of 2.17, while the value indicating the lowest recorded vulnerability was 126.16, assigned to several countries in 2012. We recoded this variable such that the lowest observed physical vulnerability values are at exactly zero, while the highest value for the most vulnerable country (that is, the Philippines in 2013) is 124.

Germanwatch acknowledges that single large-scale weather-related disasters, on which the index is based, cannot directly be attributed to anthropogenic climate change (see e.g. Kreft et al. 2016, 3). For this reason, the authors caution against using the index for far-reaching conclusions and policy decisions related to climate change vulnerability. We nevertheless include the CRI in our analysis as an alternative measure of physical vulnerability to climate change for two reasons. First, weather-related events, despite the difficulty of directly attributing them to climate change, are highly visible to policy-makers, while the long-term slow-onset effects of a changing climate are much more abstract and harder to see and understand. Some definitions of climate change adaptation, including the widely used definition by the IPCC (IPCC 2013), also specifically include adjustments to climate change and climate variability. Decisions to allocate adaptation aid might therefore often be based on more easily observable phenomena, and we include the CRI as an alternative measure of vulnerability to see if this is the case. Second, the CRI is a highly visible index and well known to policy-makers. It therefore may directly influence adaptation aid allocation decisions.

3.2.3 Adaptive Capacity and Recipient Merit

Having discussed the three indicators we use to capture physical vulnerability—that is, exposure and sensitivity to climate change—we now turn our attention to the socio-economic dimension of vulnerability—that is, adaptive capacity. Just as for vulnerability more broadly, there is no single definition of adaptive capacity (see Chap. 1). Generally speaking, adaptive capacity refers to response or coping capacity:

> It is recognized that societies adapt to a range of stimuli including, but not limited to, environmental stress. Cultures (or societies) which are able to respond to or cope with change quickly and easily are considered to have high 'adaptability' or 'capacity to adapt' (Smit and Wandel 2006, 283).

Again, adaptive capacity is difficult to measure, as it depends on many factors, including information, awareness, social cohesion, technology, and resources (see e.g. Adger et al. 2003; Barnett et al. 2008; Turner et al. 2003). However, broadly speaking, these various elements depend on two basic factors: a country's financial capacity on the one hand (especially resources and technology), and its capability in using these resources efficiently on the other (related to information, awareness, and social cohesion, among others). In order to capture adaptive capacity, we aim to operationalise both of these concepts. As discussed in Chap. 1, adaptive capacity and recipient merit are linked. Particularly the capability to use resources in an efficient manner is closely related to government efficiency and good governance, and thus to recipient merit. It is therefore inherently difficult to separate clearly adaptive capacity (H1b) from recipient merit (H2), as the variables used to operationalise these two hypotheses often encapsulate elements of both (see e.g. Weiler et al. 2017). For this reason, we discuss these variables jointly here, and include a short discussion on whether they are better able to capture adaptive capacity, recipient merit, or both.

Particularly Vulnerable Country Status

As a first—rough—measure to capture adaptive capacity, we include dummies for African countries, SIDS, and LDCS, since these groups of countries have been singled out as 'particularly vulnerable' to climate change in the climate change negotiations (e.g. UNFCCC 2009; see Chap. 2).[4] These three country groupings are not mutually exclusive, and three countries—Comoros, Guinea-Bissau, and São Tomé and Príncipe—are even members of all three categories. Among the 141 developing countries in the dataset, there are 35 SIDS, 48 LDCS, and 51 African countries.

These dummies are quite crude and capture vulnerability in a rather all-encompassing way. In other words, they mix together both physical exposure and sensitivity and adaptive capacity. For instance, while the LDC and the African dummies are good proxies for limited financial resources (and thus for low adaptive capacity),[5] they contain relatively little information on physical vulnerability. For SIDS, the opposite is true: they are considered particularly vulnerable because of their high exposure and sensitivity to climate change and other environmental changes, yet this group includes rich and poor countries with widely differing capabilities to react to these changes.[6] In other words, while most of our other measures of adaptive capacity are difficult to disentangle from recipient merit, these vulnerability dummies—particularly the SIDS dummy—are more interrelated with physical vulnerability.

GDP Per Capita

Our second measure of a country's adaptive capacity is the financial resources the government can dispose of. To operationalise financial resources we use GDP per capita figures provided by the World Bank in constant 2010 US$ (World Bank 2016). Admittedly, this variable is a somewhat rough proxy for the capacity of countries to deal with climate change, yet the advantage of GDP per capita is that data are easily available for almost all countries receiving adaptation aid. Further, many other variables suggested as measures of adaptive capacity, such as level of education or health, tend to be related to GDP per capita.

While poorer countries, less able to respond to the challenges of climate change by themselves, should receive more adaptation aid, research on development aid has found a non-linear effect of income: poorer countries receive more aid, but very poor countries receive in fact *less* than their income level would predict (e.g. Alesina and Dollar 2000; Neumayer 2003b). This non-linear relationship of poverty and aid is presumably related to countries' capacity to use resources efficiently. Donor countries want to see their funds used in a meaningful way, yet very poor countries might not have an efficient enough bureaucracy, or the elites are regarded as too corrupt and the danger of funds being embezzled too high. This argument can be tied back to the difficulty of separating adaptive capacity from good governance (that is, recipient merit), as discussed earlier. Accordingly, we expect a non-linear effect of GDP per capita. The poorest countries should receive less adaptation aid than richer ones because of their weaker governance. Yet, as income increases, adaptation aid should decrease because wealthier countries have more adaptive capacity. We are able to model both these expectations simultaneously by implementing a non-linear effect. As with other aid allocation studies (e.g. Alesina and Dollar 2000; Neumayer 2003a), we include GDP per capita in the statistical models both in linear and quadratic form. We expect a positive linear effect (in line with H1b) and a negative quadratic effect (in line with H2). The range of GDP per capita is very large, from a minimum recorded GDP per capita value in the dataset of $214 for Burundi in 2010, to a maximum recorded value of $25,335 for Equatorial Guinea, also in 2010.

Worldwide Governance Indicators

A wide range of factors play a role in determining whether countries are able to use the available resources efficiently, including how accountable governments are for their use of resources, or if (and if so, how) they control corruption. Countries that use resources more efficiently are also better able to deal with climate change—they have a higher adaptive capacity (e.g. Engle 2011; Gupta et al. 2010). Yet the efficient use of resources can also be characterised as a dimension of good governance.

The Worldwide Governance Indicators (WGIs) provided by the World Bank (Kaufmann and Kraay 2016) are a widely used measure of good governance. The WGIs encapsulate six different measures of governmental quality: Voice and Accountability, Political Stability and Absence of Violence, Government Effectiveness, Regulatory Quality, Rule of Law, and Control of Corruption (see Kaufmann et al. 2011). We believe that these variables combined are a good indication of government quality and thus useful to test our recipient merit hypothesis (H2), though the close relationship between good governance and adaptive capacity should be kept in mind.

While the six different components of the WGIs all capture a particular aspect of good governance, they are highly correlated with each other. It seems highly likely that an underlying single latent variable—good governance—is driving how well countries perform on all six WGIs. We therefore seek to combine the WGIs into one measure of good governance. To investigate whether this latent variable exists, we employ explanatory factor analysis with a single factor for these six measures. When running the analysis, we do indeed find that one single dimension is able to represent the data very well. The percentage of the explained common variance of our single factor is over 80%, and all six variables load very highly on that single factor (four factor loadings are over 0.9, one is over 0.8, and the last one is over 0.75). From this factor analysis model, we obtain the factor scores for each country, which is a single composite WGI score that indicates how well-governed a country is across the six underlying dimensions. We use this variable as a measure of good governance in our statistical models in order to test our hypothesis H2 (and to a lesser extent, hypothesis H1a). The composite WGI score ranges from −3 to 3, with higher values indicating better governance. In our dataset, the variable ranges from −2.4 for Somalia in 2010 to 1.2 for Chile in 2011.

ND-GAIN Adaptive Capacity

In Sect. 3.2.2 on physical exposure and sensitivity to climate change, we discussed the ND-GAIN. The ND-GAIN vulnerability score is composed of exposure, sensitivity, and adaptive capacity (ND-GAIN 2013). Alongside the ND-GAIN exposure sub-index, we include the ND-GAIN adaptive capacity sub-index in our analysis. Similar to the WGIs, this ND-GAIN adaptive capacity sub-index encapsulates elements of both recipient need and recipient merit. The ND-GAIN adaptive capacity sub-index considers factors such as percentage of the population with access to water and health care, maternal mortality rate, quality of the electrical grid, and quality of the trade and transport infrastructure (ND-GAIN 2013). These factors are all related to good governance—which in turn relates to adaptive capacity.

As with the ND-GAIN exposure and sensitivity sub-indices, the ND-GAIN adaptive capacity sub-index is scaled to range from 0 to 1, with higher values indicating higher vulnerability and thus lower adaptive capacity (ND-GAIN 2013). In our dataset, the ND-GAIN adaptive capacity sub-index ranges from 0.28 for Egypt in 2013 to 0.94 for Somalia in 2011. Given the way the ND-GAIN adaptive capacity sub-index is scaled, Egypt had thus the highest adaptive capacity and Somalia the lowest.

Given the links between adaptive capacity and good governance, we can formulate two opposing expectations with regard to the effect of the ND-GAIN adaptative capacity sub-index on adaptation aid: higher ND-GAIN adaptive capacity scores indicate a lower adaptive capacity as defined by ND-GAIN, and therefore a higher vulnerability to climate change. If the variable signals recipient need, we should see a ***positive*** relationship between the ND-GAIN adaptive capacity scores (lower adaptive capacity) and adaptation aid flows (H1b): countries that are more vulnerable because their adaptive capacity is lower should receive more adaptation aid. At the same time, higher ND-GAIN adaptive capacity scores also indicate low governance. In other words, the higher the score of a country, the worse its governance. If the ND-GAIN adaptive capacity sub-index signals good governance, we should see a *negative* relationship (H2): poorer governed countries should receive less adaptation aid.

3.2.4 Donor Interests

Having covered the various measures used to operationalise vulnerability, including both its physical dimension and its socio-economic dimension, and recipient merit, we now turn our attention to our final major explanatory variable for adaptation aid allocation: donor interests (H3). When donor interests are at play, it is not (only) the utility that the recipient derives from aid that explains aid allocation, but also the utility that the donor derives. This utility could be derived from economic advantages, as when the donors seeks to enable a recipient country to buy more goods (preferably from the donor). It could also be derived from political advantages, as when donors 'buy' political support in the United Nations. We use four measures that cover both donors' economic and political interests: trade volumes between a donor and a recipient; geographic distance; joint voting in the UN General Assembly; and colonial ties.

Bilateral Trade Flows

To capture donors' economic interests, we use the volume of trade between a donor and a recipient. The UN Comtrade data (UN Statistics Division 2015) reports trade of 'all commodities'. The dataset includes four trade flows for each dyad: (a) exports from country *A* to country *B* as reported by *A*; (b) imports by *A* from *B* as reported by *A*; (c) exports from *B* to *A* as reported by *B*; (d) and imports by *B* from *A* as reported by *B*. Theoretically, (a) and (d) should be identical, as exports from one country's perspective become imports from the viewpoint of the other. The same is true for (b) and (c). For highly developed country pairs, the data often are indeed very similar to each other (although they are almost never exactly the same). Yet when comparing dyads of one developed and one developing country—the dyads we are interested in—we often see a very large discrepancy between reported values. Furthermore, some developing countries do not report any trade flows at all in a non-negligible number of cases. This suggests that the statistical offices in developing countries are often not as well equipped and staffed as those in developed countries. We therefore decided to rely on trade flows as reported by developed countries (that is, by donors). The next question, then, is whether to include exports, imports, or both in the statistical analysis. Donors' economic interests lie mostly in unlocking or developing new markets for their exports, that is, in incentivising recipients to buy more of their products, not in importing more from recipients (e.g. Alesina and Dollar 2000; Clist 2011b; Dollar and Levin 2006; Younas 2008). Accordingly, we use exports from donor countries to recipients, as reported by the former. Following the donor interest model, we expect that the more donor *D* exports to recipient *R*, the more adaptation aid *R* will receive from *D*.

As is to be expected, the range of trade volumes is very high. For 1992 donor–recipient–year triplets (8.5%) in our dyadic dataset, no trade flows are reported at all, as donors do not export to all countries in the world. Two hundred and fifteen entries (or approximately 1% of all entries) record export flows from donors to recipients in excess of $10 billion. The maximum trade volume is $239 billion of exports from the USA to Mexico in 2015.

Geographic Distance

The second measure of donor interests is geographic proximity between donor and recipient country. Countries that are closer—geographically or culturally—are more present in public and media discourses within the donor (Galtung and Ruge 1965), and citizens—who pay for development aid through their taxes—may therefore expect more engagement in these countries, including through development cooperation (e.g. Dollar and Levin 2006; Hoeffler and Outram 2011). We therefore include a measure of the geographic distance between donor and recipient country to our statistical models, more specifically the minimum distance for a given dyad.

We computed this minimum distance for all dyads in the dataset using the 'cshapes' package of the R statistical computing environment (Gleditsch and Weidmann 2010). We made use of the command `distlist`, and then calculated the minimum distance (other options are available) for 1 January 2010. South Sudan did not exist and no distances were calculated for that country.[7] We use the same distance values for all years, although the minimum distance of Sudan to some donor countries after the independence of South Sudan might have slightly changed. However, this change is negligible.

In line with the donor interest model, we expect that the closer recipient *R* is to donor *D*, the more adaptation aid it should receive from *D*. Note that neighbouring countries have a distance of zero. The minimum distance in the dataset is thus zero for various country pairs (such as the USA and Mexico), while the maximum distance in the dataset is 18,916 km between South Korea and Uruguay.

Joint UN Voting

Our third measure of donor interest is voting behaviour in the UN General Assembly. The more often two countries cast the same vote (e.g., 'yes'), the closer they are politically. The UN General Assembly Voting Data measures to what extent two countries have common interests in the international sphere (Voeten et al. 2009). More specifically, we use the dyadic affinity score provided by the dataset (called 's2un' in the dataset's codebook), which measures how often dyad members jointly vote 'yes' or 'no' in the UN General Assembly. The calculated raw score is then standardised in a second step to range from −1 to +1. If a country pair agrees on all issues, that is, if the two countries of the pair vote in the same way on all votes in a given year, they obtain the maximum possible value of 1. If they always disagree, the minimum value of −1 is assigned.

In line with the donor interest model we expect higher affinity scores to translate to higher adaptation aid flows. In other words, the more often recipient *R* casts the same vote as donor *D*, the more adaptation aid *R* should receive from *D*. In our dataset, the affinity scores include the full theoretical range. Three country pairs—São Tomé and Príncipe and the USA in 2010; Equatorial Guinea and the USA in 2012; and Kiribati and the USA in 2013—never voted in the same way; 1759 country pairs always voted in the same way.

Colonial Ties

As our fourth and last measure of donor interest, we include colonial ties. These measure historic relations, present day political interests, as well as (to some degree) cultural similarities. We use a dummy variable that records which Western country was the colonial power at the time of independence.

The data on colonial relationships between two countries are from Hadenius and Teorell (2007), as reported in the Quality of Government Institute Dataset (Teorell et al. 2015). The variable only records Western overseas colonialism and excludes settler colonies as well as non-Western colonisers. If a country had more than one colonial power, only the last one is recorded in the dataset. Note that not all recipient countries have been colonies; if a recipient country has never experienced colonisation, no ties are recorded. Nor did all donors have colonies. In our dataset, only 8 of the 28 donors had colonies (Belgium, France, Italy, Netherlands, Portugal, Spain, the UK, and the USA), the remaining 20 donors do not have any entries on this variable. Only one of the recipient countries was missing in this dataset: Timor Leste. We therefore manually added this entry in our final dataset, with Portugal being the former colonial power. We expect that a donor will provide more adaptation aid to its former colonies.

3.2.5 Control Variables

In addition to all the variables capturing vulnerability to climate change, recipient merit, and donor interest, we also include two control variables in the statistical analysis: population size and total development aid.

Larger countries—in terms of their population—are more likely to receive at least some aid from all donors because of their larger economic and political weight. To some extent, population size is thus a measure of donor interests. Yet larger countries at the same time receive less aid per capita. In terms of per capita aid, the aid allocation literature has clearly documented a 'small country bias'. Smaller countries are likely to receive more aid than large countries on a per capita basis (e.g. Alesina and Dollar 2000; Neumayer 2003a; Younas 2008). For instance, in 2012 Tanzania was the largest beneficiary of German adaptation aid, receiving over $73 million for principal and significant adaptation projects. Given Tanzania's population of over 45 million, this only translated to about $1.46 per capita. In contrast, Mongolia, which is much smaller with a population of less than 3 million, received about $15 million in adaptation aid from Germany in 2012, or $5.34 per capita. Including population helps us to estimate the effects of our independent variables separately from population size effects. Data for the total population of a country comes from the World Bank (World Bank 2016), and ranges from just under 10,000 for Tuvalu[8] to 1.4 billion for China.

The second control variable we employ is total development aid. Adaptation aid is a subset of total development aid, and allocation decisions for development aid and adaptation aid are interdependent, particularly since adaptation is increasingly 'mainstreamed' into standard aid projects (see also Chap. 6). Accordingly, it is very likely that adaptation aid is closely linked with and correlated to development aid, which makes controlling for total development aid essential. Data on overall development aid are taken from the OECD CRS, as for the dependent variables of this study. Again, we use dyadic data which record the total amount of bilateral aid flows from a given donor to a given recipient in a given year. Since not all donors provide aid to all recipients, the lowest value of total development aid is zero. The maximum value is $6.2 billion, which the United Arab Emirates provided to Egypt in 2013.

Table 3.1 below lists all variables and sources from which we took the data. Table A.1 in the Appendix provides summary statistics for all variables used in the statistical analysis.

3.2.6 *Modelling Strategy*

Now that we have covered our dependent and independent variables, we can explain our modelling strategy for both the pooled models including all donors, and the separate models for the three selected countries, that is, for Germany, Sweden, and the UK. To be able to run regression models, we compiled all variables described in the previous sections into a single dataset.

In total, we have 28 donor and 141 recipient countries in the final dataset, and data on adaptation aid for 6 years from 2010 through to 2015. Of the 28 donor countries, 26 are present in the dataset for the entire time period of the study; two donors (the Czech Republic and Iceland) only started to provide adaptation aid in 2011 and hence are only covered for 5 years. Two donors (Poland and the Slovak Republic) started to provide aid (both development aid and adaptation aid) even later, in 2013 (see OECD 2016b). Given the short time span and the relatively small sums involved, we consider these two countries still to be in a learning and adjusting stage and did not include them in the analysis. Two non-OECD countries have reported aid figures constantly since 2010: Kuwait and the United Arab Emirates. We therefore include these two donors in our dataset and the analysis.

For the 2011–2015 period, 141 countries are listed on the OECD DAC list of eligible countries. In 2010, five more countries were still eligible to receive aid: Barbados, Croatia, Mayotte, Oman, and Trinidad and Tobago. These countries were removed from the OECD DAC List of ODA Recipients in 2011, and therefore only received adaptation aid in 2010. We excluded them from the analysis. The remaining 141 recipients are in the dataset for the entire time period of the study. Overall, the number of observations in the full dataset is 23,406.[9] However, due to missing values in the covariates, the number of observations in the various models reported in the next chapter is somewhat lower than that figure, and varies depending on which variables are included.

Following suggestions from the aid allocation literature, we employ a two-stage Cragg Model (Clist 2011a,b; Manning et al. 1987), which allows a separation of aid allocation into a selection and an allocation stage. Donors decide first to which countries they want to give some aid (the selection stage) and in a second stage (the allocation stage), how much aid they want to give to the selected countries (e.g. Clist 2011b).[10] The two stages are econometrically modelled separately, but the allocation stage must be interpreted conditionally on receiving adaptation aid at the selection stage (see Clist 2011b).

In addition, we include at both stages donor fixed effects, for a donor's adaptation aid allocation decisions in a given year cannot be regarded as entirely independent. As the overall budget for adaptation aid is limited in a given year, when the donor decides to give a certain amount of aid to recipient *R*, this reduces the amount of aid that is still available for recipient *S*. Finally, we also include year-fixed effects to account for path dependencies: if a donor has given adaptation aid to recipient *R* in the past, the donor may be more (or less) likely to allocate aid to that recipient (see also Chaps. 2 and 6).

We run four separate models. First, we test the role of physical vulnerability to climate change (H1a). The first model specification uses only the three measures of physical vulnerability—the ND-GAIN exposure sub-index, EVI, and CRI. The second model specification combines our measures of adaptive capacity (H1b) and recipient merit (H2), as it is difficult to separate these two effects. The second model thus includes the dummies for vulnerable country status, the ND-GAIN adaptive capacity sub-index, GDP per capita in its linear and quadratic form, as well as the composite WGI index. The third model specification includes only the four measures of donor interests (H3)—exports, geographic distance, colonial dummy, and joint voting in the UN General Assembly. Our fourth and final model specification combines all the variables for all hypotheses. The two control variables of population size and total development aid are included in all these model specifications.

In sum, we have four model specifications, two dependent variables—one using principal adaptation aid only and one using principal and discounted significant adaptation aid—and two stages of adaptation aid allocation—the selection and the allocation stage. In total, we have 16 regression models for all donors combined. Additionally, we run the two-stage regression models for our three case study countries—Germany, Sweden, and the UK—separately. Chapter 5 describes all models and our interpretation of all results.

Finally, for each variable described above we had to decide how to use them in our models. First, it is standard practice to use the logarithmic transform for highly skewed variables. Such a transformation converts these variables so that they more closely resemble a normal distribution, which in turn helps to avoid violating the basic assumptions of regression analysis. Second, many of our independent variables, in theory, inform policy-makers in donor countries when making decisions—for instance about the vulnerability of recipients. However, when they base decisions on such information, there will be a certain time lag between the information being available and the decision to act on it. We therefore lag all time-variant independent variables by 1 year, to allow for the time needed before the information the variable contains becoming available to policy-makers. Table 3.1 summarises all variables and indicates whether they are lagged and/or logarithmised in the statistical analysis.

3.3 Qualitative Design

Let us turn to the qualitative part of the analysis, which, in combination with the large-n statistical analysis, helps us understand adaptation aid allocation more deeply and more comprehensively. Our qualitative analysis is based on semi-structured interviews and key policy documents, which provide thick and detailed descriptions and add explanations and nuance to the patterns observed in the statistical analysis, thus increasing the validity of our findings (Lynch 2013; Martin 2013).

We have already discussed the case selection criteria and our most different systems design in Sect. 3.1, based on which we selected Germany, Sweden, and the UK for in-depth analysis. As described in the previous section, we run separate regression models for these three selected cases. In addition to this statistical analysis, we also qualitatively explore the role of climate change and adaptation in development cooperation in these countries, based on semi-structured in-depth interviews and key policy documents. We conclude the methods chapter with a discussion of our

Table 3.1 Summary of all variables used in the statistical analysis

Indicator and source	Lagged	Logged
Dependent variables: adaptation aid		
– Principal adaptation aid, per capita (OECD 2016b)		✓
– Principal and discounted significant adaptation aid, per capita (OECD 2016b)		✓
Independent variables		
Recipient need: exposure and sensitivity (H1a)		
– ND-GAIN exposure sub-index (ND-GAIN n.d.)	✓	
– Environmental Vulnerability Index (Kaly et al. 2004)	✓	
– Climate Risk Index (Germanwatch n.d.)	✓	
Recipient need: adaptive capacity (H1b)		
– Vulnerable country dummies (LDCs, SIDS, African countries)		
– GDP per capita (World Bank 2016)	✓	✓
Recipient merit (H2)		
– Efficient resource use (WGI composite index) (Kaufmann and Kraay 2016)	✓	
– ND-GAIN adaptive capacity sub-index (ND-GAIN n.d.)	✓	
Donor interests (H3)		
– Exports from donors to recipients (UN Statistics Division 2015)	✓	✓
– Joint UN General Assembly voting (Voeten et al. 2009)	✓	
– Geographic distance (Gleditsch and Weidmann 2010)		✓
– Colonial ties (Teorell et al. 2015)		
Control variables		
– Total population (World Bank 2016)	✓	✓
– Total development aid (OECD 2016b)	✓	✓

qualitative empirical analysis. We explain how we selected our interview partners and what questions we asked, which policy documents we used, as well as how we analysed the interview and document data.

3.3.1 Semi-structured Interviews

For our interviews, we targeted policy-makers, practitioners, and observers working in the aid sector in the three selected countries. Specifically, we contacted staff in the development, environment, and foreign affairs ministries, as well as in the different aid agencies and members of parliament active in development committees. Beyond these government officials, we contacted independent experts working on adaptation aid, notably observers from NGOs and think tanks. We also used snowball sampling as we asked the people we contacted for suggestions for further interviewees, and were partly referred to colleagues and contacts within the same organisation or elsewhere. In total, we contacted 15 organisations in Germany and talked to 6; we contacted 11 organisations in Sweden and talked to 8; and we contacted 13 organisations in the UK and talked to 7. The exact number of interviews, however, slightly differs, as we twice talked to 2 individuals from the same organisation (in Germany). Overall, we conducted 28 interviews between March and June 2016. Most interviews were face-to-face, but in some cases, we conducted interviews by phone or via Skype. In Germany, we conducted the interviews in German, and in English in Sweden and the UK. We audio-recorded almost all interviews as well as took handwritten notes. We transcribed the audio-records and used the transcripts for analysis (see Sect. 3.3.3 below). Table 3.2 lists interviews by country and type of interviewee (government or observer). The identifiers in column 1 of Table 3.2 correspond to the identifiers used in Chap. 6.

The interviews were designed as semi-structured expert interviews (Liebold and Trinczek 2009; Meuser and Nagel 2009). Our questionnaire, reported in Table A.2 in the Appendix, asked for: information about the interview partner and the organisation they worked for; the role of climate change and specifically adaptation in development work; the decision-making process behind (adaptation) aid allocation; reporting adaptation aid; and future developments. These questions were intended to guide the conversation, but we did not strictly adhere to these specific questions or the order in which we asked them.

3.3.2 Policy Documents

Key policy documents served as a second source of information alongside the semi-structured interviews. We started with the OECD peer reviews. Members of the OECD DAC regularly examine the development policies and performance of their peers and publish the results of this examination in the so-called peer review reports. Starting from the OECD peer reviews for Germany (OECD 2015), Sweden, (OECD 2013) and the UK (OECD 2014), we identified central policy documents and information brochures in the three countries that lay out the broad frameworks of development

Table 3.2 List of interviews conducted in Sweden, Germany, and the UK

#	Interview with	Date
Germany		
Government		
DE1[a]	Federal Ministry for Economic Cooperation and Development (BMZ), Special Unit 'Climate'	14/4/2016
DE2	Federal Ministry for Economic Cooperation and Development (BMZ), OECD/DAC; ODA statistics division	25/5/2016
DE3	KfW Development Bank, Environment and Climate Competence Centre	15/3/2016
DE4[b]	Deutsche Gesellschaft für Internationale Zusammenarbeit (GIZ)	17/3/2016
DE5	Member of Parliament for the Conservatives (CDU)	25/5/2016
DE6	Member of Parliament for the Green Party (Bündnis 90/Die Grünen)	26/5/2016
Observers		
DE7	Deutsches Institut für Entwicklung/German Development Institute (DIE)	14/3/2016
DE8	German Institute for Development Evaluation (DEval; formerly DIE)	17/3/2016
DE9	Germanwatch	14/3/2016
DE10[a]	Germanwatch	27/4/2016
DE11[a]	Independent Consultant (deutscheklimafinanzierung.de)	7/6/2016
DE12[a]	Oxfam Germany	8/6/2016
Sweden		
Government		
SW1	Swedish Ministry for Foreign Affairs, Department for Multilateral Development Cooperation	30/5/2016
SW2	Swedish International Development Cooperation Agency (Sida)	31/5/2016
SW3	Swedish Ministry of the Environment and Energy, Climate Division	2/6/2016
Observers		
SW4	Stockholm Environment Institute	30/5/2016
SW5	Diakonia	31/5/2016
SW6	We Effect	31/5/2016
SW7	Church of Sweden international/Svenska kyrkans internationella arbete	1/6/2016
SW8	Naturskyddsföreningen (Swedish Society for Nature Conservation)	2/6/2016

(continued)

Table 3.2 (continued)

United Kingdom		
Government		
UK1	Department for International Development (DfID), Research and Evidence Division	9/5/2016
Observers		
UK2[a]	International Centre for Climate Change and Development, Independent University, Bangladesh	13/4/2016
UK3	Overseas Development Institute (ODI)	7/6/2016
UK4	Care International	9/5/2016
UK5	Climate and Development Knowledge Network (CDKN)	10/5/2016
UK6	Oxfam	11/5/2016
UK7[a]	Independent Climate Finance Researcher/Consultant	13/6/2016

Note: All information from our interviews represent the personal experiences and views of the interviewees and not necessarily of the organisation they work for
[a] Conducted by phone or via Skype
[b] Not audio-recorded

cooperation as well as provide specific information and guidelines on environmental, climate change, and adaptation questions. We focused on documents published between 2010 and 2016—roughly the same period for which we have quantitative data. Additionally, we searched for political documents: party programmes, coalition agreements, statements of government policy and position papers of political parties in power. Overall, this yielded 11 documents for Germany, 14 documents for Sweden, and 10 documents for the UK. Table 3.3 lists all documents that we analysed.

3.3.3 *Analysing Our Interview and Document Data*

Once we had assembled our qualitative data, we used NVivo to analyse it in three steps: we first assigned thematic codes to the text, then structured or clustered the coded material, and finally synthesised it (see e.g. Meuser and Nagel 2009; Richards 2014).

In a first step, we read through all the interview transcripts as well as the policy documents and assigned thematic codes to relevant text passages. Some of these thematic codes were already given by our research question and interview guide, such as recipient need, recipient merit, and donor interests. We created additional codes as necessary and regrouped

Table 3.3 Overview of the policy documents used in the qualitative analysis

#	Type of policy document	Reference
Germany		
1	OECD Peer Review: Germany 2015	OECD (2015)
2	Coalition Agreement between Conservative Party and Liberal Party, 17th election period (2009–2013)	Government of Germany (2009)
3	Coalition Agreement between Conservative Party and Social Democratic Party, 18th election period (2013–2017)	Government of Germany (2013)
4	position paper on development policy of the Conservative Party	Fraktion im Bundestag (2013)
5	position paper on development policy of the Social Democratic Party	Raabe et al. (2013)
6	booklet by the GIZ on adaptation	Olivier et al. (2013)
7	14th Report of the Federal Government on Development Policy	BMZ (2013)
8	German Aid Policy Framework (Charter for the Future)	BMZ (2015a)
9	Booklet by the Federal Ministry for Economic Cooperation and Development on Climate Action	BMZ (2015b)
10	Booklet by the Federal Ministries for the Environment, Nature Conservation and Nuclear Safety and for Economic Cooperation and Development	BMU and BMZ (2013)
11	Information Brochure 1/2011 by the Federal Ministry for Economic Cooperation and Development on Climate Change and Development	BMZ (2011)
Sweden		
1	OECD Peer Review: Sweden 2013	OECD (2013)
2	2010 Party programme of the centre-right Allians Party	Alliansen (2010)
3	Coalition Agreement between Social Democratic Party and Green Party (2014–2018)	Government of Sweden (2014b)
4	Statements of Government Policy, 2010–2015	Government of Sweden (2010, 2011, 2012, 2013, 2014c, 2015)
5	Government Bill: Sweden's Policy for Global Development	Government of Sweden (2003)
6	Government Communication: Aid Policy Framework 2013	Government of Sweden (2014a)
7	Government Communication: Aid Policy Framework 2016	Government of Sweden (2016)
8	Policy for Environmental and Climate Issues in Swedish Development Cooperation, 2010–2014	Government Offices of Sweden (2010)
9	Booklet by Sida on the Special Climate Change Initiative	Wasielewski Ahlfors (2011)

(continued)

Table 3.3 (continued)

United Kingdom		
1	OECD Peer Review: United Kingdom 2014	OECD (2014)
2	2010 Party programme of the Conservative Party	Conservative Party (2010)
3	2015 Party programme of the Conservative Party	Conservative Party (2015)
4	2010 Party programme of the Liberal Democrats	Liberal Democrats (2010)
5	Coalition Agreement between Conservative Party and Liberal Democrats (2010–2015)	Government of the United Kingdom (2010)
6	Parliament Act on Official Development Assistance Target	Parliament of the United Kingdom (2015)
7	DfID Business Plan 2011–2015	DfID (2011b)
8	DfID Annual Report 2014–2015	Df ID (2015)
9	Bilateral Aid Review	DfID (2011a)
10	Booklet by DfID on work on the environment	DfID (2012)

them, turning codes into sub-codes or aggregating sub-codes into codes (Richards 2014). Our final coding scheme consisted of 11 different codes: overall policies and visions; adaptation versus development; comparison across donors; decision-making process; recipient need; recipient merit; donor interests; finance modalities; fairness; UNFCCC negotiations; and future/outlook. Some of these codes had sub-codes; recipient merit for instance includes two aspects, good governance and climate commitment (see Chap. 6). Text passages could be assigned more than one thematic code. To illustrate, let us look at the following statement from the OECD peer review for Sweden: 'in making its choices it prioritises poor people in low income countries and also countries that are struggling with good governance and human rights' (OECD 2013, 41). This sentence was assigned two codes, one for poverty, a sub-code of recipient need, and one for good governance, a sub-code of recipient merit.

Once we had coded all interview transcripts and policy documents, we structured or clustered the coded material (Liebold and Trinczek 2009; Richards 2014). Again, our research question had already provided us with a structure to reorganise and synthesise the data, namely the aid allocation process with a focus on the three determinants of aid allocation (recipient need, recipient merit, and donor interests). We identified two additional overarching clusters or categories that came out of the material: the overall role of climate change and adaptation in development cooperation as well as issues and challenges, including the additionality of adaptation finance, reporting and accounting for adaptation finance, and the level of adaptation finance. We read through all passages that were assigned a specific thematic code and organised the material by cluster. Within each cluster, we identified topics and points. For example, under the category 'overall role of climate change and adaptation in development cooperation', interview partners and policy documents emphasised that climate change is an important priority area and that its importance has increased over recent years. We listed relevant quotes and statements under these two points.

In a final step, we synthesised the information and summarised points made across interviews and policy documents as well as differences between them—this is what we report in Chap. 6. Note that we used all material in its original language for the first two steps and only translated quotes at this third step.

In sum, this chapter has laid out our research design. Our empirical analysis proceeds in three steps. We start with a broad overview of adaptation aid flows, as reported in the OECD CRS in the next chapter. We then test statistically how donors allocate their adaptation aid across recipients, and to what extent their decisions are guided by recipient need, recipient merit, and donor interests. In a third step, we zoom in on three large adaptation donors, Germany, Sweden, and the UK. We first assess the distribution of their adaptation aid statistically and then explore the decision-making processes qualitatively. In this chapter, we have explained why we opted for such a mixed methods research design, described which variables we use in our statistical analysis, and discussed how we collected and analysed the material for our qualitative case studies. Now we can proceed to the findings of our analysis.

Notes

1. For a detailed comparison of aid flows (both development aid and adaptation aid) in the three selected country cases, but also across the wider donor landscape, see the descriptive statistics in Chap. 4.
2. In doing so, we generate a dataset with many empty entries when adaptation aid flows between a donor and a recipient in a given year are zero. This allows us to model two types of decisions: first, whether a donor provides any adaptation aid to a specific recipient in a given year (selection stage), and second, if so, how much the donor provides to that selected recipient

(allocation stage). We will discuss the modelling strategy later in this chapter.

3. All reports and data are available from Germanwatch's website at http://germanwatch.org/en/cri.
4. The United Nations lists countries that are LDCs and/or SIDS, see http://www.un.org/en/development/desa/policy/cdp/ldc/ldc_list.pdf and http://unohrlls.org/about-sids/country-profiles/.
5. Although of course GNI per capita is rather high for some African countries like Seychelles.
6. While several SIDS are also LDCs, other SIDS are no longer eligible for ODA because their GNI per capita is by now too high. Anguilla and St Kitts and Nevis, for instance, were removed from the list in 2014 (see http://www.oecd.org/dac/stats/historyofdaclistsofaidrecipientcountries.htm#Chronology).
7. Since many other variables were not available for South Sudan either, we excluded the country from our dyadic dataset.
8. The smallest recipient country in the dataset is Niue with a population of only 1500. However, Niue is not included in the World Bank data. Since this and most other variables are not available for Niue, the statistical analysis in Chap. 5 does not include Niue.
9. Each recipient is in the dataset 166 times, as each country forms a dyad with 26 donors 6 times, and with the Czech Republic and Iceland 5 times ($26 \times 6 + 2 \times 5 = 166$). Thus, over all the 141 recipient countries, we obtain $166 \times 141 = 23{,}406$ observations.
10. There is another stage in the decision-making process of how to allocate aid, namely whether to provide multilateral or bilateral aid. This first step is considered indirectly in our research, as we take into account only those adaptation aid flows which donors decided to provide bilaterally. Note that the three stages of the decision-making process occur at different points in time and that circumstances may have changed between these stages.

References

Adaptation Watch. 2015. Toward Mutual Accountability: The 2015 Adaptation Finance Transparency Gap Report. Policy Briefing. Available online at http://www.adaptationwatch.org/.

Adger, W.N., S. Huq, K. Brown, D. Conway, and M. Hulme. 2003. Adaptation to Climate Change in the Developing World. *Progress in Development Studies* 3 (3): 179–195.

Alesina, A. and D. Dollar. 2000. Who Gives Foreign Aid to Whom and Why? *Journal of Economic Growth* 5: 33–63.

Alliansen. 2010, August. Jobbmanifestet: Alliansens Valmanifest 2010–2014. Stockholm.

Anckar, C. 2008. On the Applicability of the Most Similar Systems Design and the Most Different Systems Design in Comparative Research. *International Journal of Social Research Methodology* 11 (5): 389–401.

Anemüller, S., S. Monreal, and C. Bals. 2006. Global Climate Risk Index 2006. Weather Related Loss Events and Their Impact on Countries in 2004 and in a Long-Term Comparison. Available online at https://germanwatch.org/de/download/13503.pdf.

Barnett, J., S. Lambert, and I. Fry. 2008. The Hazards of Indicators: Insights from the Environmental Vulnerability Index. *Annals of the Association of American Geographers* 98 (1): 102–119.

BMU and BMZ. 2013. Together for a Common Cause: Germany's Contribution to International Climate Financing. Bundesministerium für Umwelt, Naturschutz, Bau und Reaktorsicherheit and Bundesministerium für wirtschaftliche Zusammenarbeit und Entwicklung, Berlin.

BMZ. 2011, January. Klimawandel und Entwicklung. Information Brochure 1/2011 of the Federal Ministry for Economic Cooperation and Development, Berlin.

BMZ. 2013, May. 14. Entwicklungspolitischer Bericht der Bundesregierung. Weißbuch zur Entwicklungspolitik. Federal Ministry for Economic Cooperation and Development (BMZ), Berlin.

BMZ. 2015a. Charter for the Future: One World—Our Responsibility. Federal Ministry for Economic Cooperation and Development (BMZ), Bonn.

BMZ. 2015b, April. Climate Action in Practice. The Contribution of German Development Policy. Federal Ministry for Economic Cooperation and Development (BMZ), Berlin.

Brooks, N., W.N. Adger, and P.M. Kelly. 2005. The Determinants of Vulnerability and Adaptive Capacity at the National Level and the Implications for Adaptation. *Global Environmental Change* 15: 151–163.

Brown, J., N. Bird, and L. Schalatek. 2010, June. Climate Finance Additionality: Emerging Definitions and Their Implications. Climate Finance Policy Brief No. 2. Heinrich Böll Stiftung. The Green Political Foundation. Available online at https://www.odi.org/sites/odi.org.uk/files/odi-assets/publications-opinion-files/6032.pdf.

CDU-CSU Fraktion im Bundestag. 2013, April. Entwicklungspolitisches Positionspapier. Arbeitsgruppe Wirtschaftliche Zusammenarbeit und Entwicklung, Berlin.

Clist, P. 2011a. 25 Years of Aid Allocation Practice: Comparing Donors and Eras. Nottingham: Centre for Research in Economic Development and International Trade, University of Nottingham (Discussion Papers 9/2011).

Clist, P. 2011b. 25 Years of Aid Allocation Practice: Whither Selectivity? *World Development* 39 (10): 1724–1734.

Conservative Party. 2010, April. *Invitation to Join the Government of Britain: The Conservative Manifesto 2010.* London: Conservative Campaign Headquarters.

Conservative Party. 2015. *Strong Leadership: A Clear Economic Plan. A Brighter, More Secure Future.* London: Conservative Campaign Headquarters.

Creswell, J.W. 2013. *Research Design: Qualitative, Quantitative, and Mixed Methods Approaches.* Thousand Oaks: Sage Publications.

DfID. 2011a, March. Bilateral Aid Review: Technical Report. Department for International Development, London.

DfID. 2011b, May. Business Plan 2011–2015, Department for International Development. Department for International Development, London.

DfID. 2012, June. Supporting a Healthy Environment: A Fresh Approach to Our Work on the Environment. Department for International Development, London.

DfID. 2015, July. Annual Report and Accounts 2014–2015. Department for International Development, London.

Dollar, D., and V. Levin. 2006. The Increasing Selectivity of Foreign Aid, 1984–2003. *World Development* 34 (12): 2034–2046.

Donner, S.D., M. Kandlikar, and S. Webber. 2016. Measuring and Tracking the Flow of Climate Change Adaptation Aid to the Developing World. *Environmental Research Letters* 11 (5): 1–9.

Engle, N.L. 2011. Adaptive Capacity and Its Assessment. *Global Environmental Change* 21: 647–656.

Francke Lund, H., C. Clapp, A. Torvanger, M.S. Jane Wilkinson, Barbara Buchner, F. Mazza, P. Oliver, and D. Abramskiehn. 2015. Background Report on Long-term Climate Finance prepared for the German G7 Presidency 2015. Technical Report, CICERO and Climate Policy Initiative, Eschborn.

Galtung, J., and M.H. Ruge. 1965. The Structure of Foreign News: The Presentation of the Congo, Cuba and Cyprus Crises in Four Norwegian Newspapers. *Journal of Peace Research* 2 (1): 64–90.

Germanwatch (n.d.). Global Climate Risk Index. Available online at http://germanwatch.org/en/cri.

Gleditsch, K., and N. Weidmann. 2010. Mapping and Measuring Country Shapes: The Cshapes Package. *R Journal* 2 (1): 18–23.

Government of Germany. 2009. Wachstum. Bildung. Zusammenhalt. Koalitionsvertrag zwischen CDU, CSU und FDP, 17. Legislaturperiode. Available online at https://www.bmi.bund.de/SharedDocs/Downloads/DE/Ministerium/koalitionsvertrag.pdf.

Government of Germany. 2013. Deutschlands Zukunft gestalten. Koalitionsvertrag zwischen CDU, CSU und SPD, 18. Legislaturperiode. Available online at https://www.bundesregierung.de/Content/DE/_Anlagen/2013/2013-12-17-koalitionsvertrag.pdf.

Government of Sweden. 2003, May. Shared Responsibility: Sweden's Policy for Global Development. Government Bill 2002/03:122, Stockholm.

Government of Sweden. 2010, October. Regeringsförklaringen 2010. Available online at http://www.regeringen.se/informationsmaterial/2010/10/regeringsforklaringen-5-oktober-2010/, Stockholm.

Government of Sweden. 2011, September. Regeringsförklaringen 2011. Available online at http://www.regeringen.se/informationsmaterial/2011/09/regeringsforklaringen-15-september-2011/, Stockholm.

Government of Sweden. 2012, September. Regeringsförklaringen 2012. Available online at http://www.regeringen.se/informationsmaterial/2012/09/regeringsforklaringen-18-september-2012/, Stockholm.

Government of Sweden. 2013, September. Regeringsförklaringen 2013. Available online at http://www.regeringen.se/informationsmaterial/2013/09/regeringsforklaringen-17-september-2013/, Stockholm.

Government of Sweden. 2014a, March. Aid Policy Framework – The Direction of Swedish Aid. Government Communication 2013/14:131.

Government of Sweden. 2014b. Överenskommelse mellan Socialdemokraterna och Miljöpartiet de gröna. Available online at http://www.socialdemokraterna.se/upload/Overenskommelse_mellan_Socialdemokraterna_och_Miljopartiet_de_grona.pdf, Stockholm.

Government of Sweden. 2014c, October. Regeringsförklaringen 2014. Available online at http://www.regeringen.se/informationsmaterial/2014/10/regeringsforklaringen-3-oktober-2014/, Stockholm.

Government of Sweden. 2015, September. Regeringsförklaringen 2015. Available online at http://www.regeringen.se/tal/2015/09/regeringsforklaringen-den-15-september-2015/, Stockholm.

Government of Sweden. 2016, December. Policyramverk för svenskt utvecklingssamarbete och humanitärt bistånd. Regeringens skrivelse 2016/17:60, Stockholm.

Government of the United Kingdom. 2010, May. *The Coalition: Our Programme for Government. Freedom. Fairness. Responsibility.* London: Cabinet Office.

Government Offices of Sweden. 2010. Policy for Environmental and Climate Issues in Swedish Development Cooperation, 2010–2014. Ministry of Foreign Affairs Sweden, Stockholm.

Guillaumont, P., and C. Simonet. 2011, March. Designing an Index of Physical Vulnerability to Climate Change. Available online at http://www.ferdi.fr/sites/www.ferdi.fr/files/idi/B18.pdf. Ferdi Policy Brief.

Gupta, J., C. Termeer, J. Klostermann, S. Meijerink, M. van den Brink, P. Jong, S. Nooteboom, and E. Bergsma. 2010. The Adaptive Capacity Wheel: A

Method to Assess the Inherent Characteristics of Institutions to Enable the Adaptive Capacity of Society. *Environmental Science & Policy 13* (6): 459–471.

Hadenius, A., and J. Teorell. 2007. Pathways from Authoritarianism. *Journal of Democracy 18* (1): 143–157.

Hinkel, J. 2011. Indicators of Vulnerability and Adaptive Capacity: Towards a Clarification of the Science–Policy Interface. *Global Environmental Change 21*: 198–208.

Hoeffler, A., and V. Outram. 2011. Need, Merit, or Self-Interest—What Determines the Allocation of Aid? *Review of Development Economics 15* (2): 237–250.

IPCC. 2013. Annex III: Glossary. In *Climate Change 2013: The Physical Science Basis. Contribution of Working Group I to the Fifth Assessment Report of the Intergovernmental Panel on Climate Change*, ed. T. Stocker, D. Qin, G.-K. Plattner, M. Tignor, S. Allen, J. Boschung, A. Nauels, Y. Xia, V. Bex, and P. Midgley, 1447–1466. Cambridge and New York: Cambridge University Press.

Junghans, L., and S. Harmeling. 2012. Different Tales from Different Countries: A First Assessment of the OECD 'Adaptation Marker'. Germanwatch Briefing Paper.

Kaly, U., C. Pratt, and J. Mitchell. 2004. The Environmental Vulnerability Index (EVI) 2004. Secretariat of the Pacific Community Applied Geoscience and Technology Division (SOPAC) Technical Report 384. Available online at http://www.sopac.org/sopac/evi/Files/EVI%202004%20Technical%20Report.pdf.

Kaufmann, D., and A. Kraay. 2016. Worldwide Governance Indicators. Available online at http://info.worldbank.org/governance/wgi/index.aspx#home.

Kaufmann, D., A. Kraay, and M. Mastruzzi. 2011. The Worldwide Governance Indicators: Methodology and Analytical Issues. *Hague Journal on the Rule of Law 3* (2): 220–246.

Kreft, S., and D. Eckstein. 2014. Global Climate Risk Index 2014. Who Suffers Most from Extreme Weather Events? Weather-Related Loss Events in 2012 and 1993 to 2012. Germanwatch Briefing Paper. Available online at www.germanwatch.org/en/cri.

Kreft, S., D. Eckstein, L. Dorsch, and L. Fischer. 2016. Global Climate Risk Index 2016. Who Suffers Most From Extreme Weather Events? Weather-Related Loss Events in 2014 and 1995 to 2014 . Available online at https://germanwatch.org/de/download/13503.pdf.

Liberal Democrats. 2010. Liberal Democrat Manifesto 2010. Liberal Democrats, London.

Liebold, R., and R. Trinczek. 2009. Experteninterview. In *Handbuch Methoden der Organisationsforschung: Quantitative und Qualitative Methoden*, ed. S. Kühl, P. Strodtholz, and A. Taffertshofer, 32–56. Wiesbaden: VS Verlag für Sozialwissenschaften.

Lynch, J.F. 2013. Aligning Sampling Strategies with Analytic Goals. In *Interview Research in Political Science*, Chap. 1, ed. L. Mosley, 31–44. Ithaca and London: Cornell University Press.

Manning, W., N. Duan, and W. Rogers. 1987. Monte Carlo Evidence on the Choice Between Sample Selection and Two-Part Models. *Journal of Econometrics 35* (1): 59–82.

Martin, C.J. 2013. Crafting Interviews to Capture Cause and Effect. In *Interview Research in Political Science*, Chap. 5, ed. L. Mosley, 109–123. Ithaca and London: Cornell University Press.

Meuser, M., and U. Nagel. 2009. Das Experteninterview – konzeptionelle Grundlagen und methodische Anlage. In *Methoden der vergleichenden Politik- und Sozialwissenschaft: neue Entwicklungen und Anwendungen*, ed. S. Pickel, G. Pickel, H.-J. Lauth, and D. Jahn, 465–579. Wiesbaden: VS Verlag für Sozialwissenschaften.

Michaelowa, A., and K. Michaelowa. 2011. Coding Error or Statistical Embellishment? The Political Economy of Reporting Climate Aid. *World Development* 39 (11): 2010–2020.

ND-GAIN. 2013, December. University of Notre Dame Global Adaptation Index Detailed Methodology Report. Available online at http://www3.nd.edu/~nchawla/methodology.pdf. Last accessed 15 March 2017.

ND-GAIN (n.d.). 2017. ND-GAIN: Notre Dame Global Adaptation Index. Available online at http://index.gain.org/. Last accessed 15 March 2017.

Neumayer, E. 2003a. *The Pattern of Aid Giving: The Impact of Good Governance on Development Assistance.* London: Routledge.

Neumayer, E. 2003b. The Determinants of Aid Allocation by Regional Multilateral Development Banks and United Nations Agencies. *International Studies Quarterly* 47 (1): 101–122.

Newman, I., and C.R. Benz. 1998. *Qualitative-Quantitative Research Methodology: Exploring the Interactive Continuum.* Carbondale: SIU Press.

OECD. 2011. Handbook on the OECD-DAC Climate Markers. Available online at www.oecd.org/dac/stats/48785310.pdf.

OECD. 2013. oecd *Development Co-operation Peer Review: Sweden 2013.* Paris: OECD Publishing.

OECD. 2014. oecd *Development Co-operation Peer Reviews: United Kingdom 2014.* Paris: OECD Publishing.

OECD. 2015. oecd *Development Co-operation Peer Reviews: Germany 2015.* Paris: OECD Publishing.

OECD. 2016a. Development Aid in 2015 Continues to Grow Despite Costs for In-donor Refugees. Available online at http://www.oecd.org/dac/stats/ODA-2015-detailed-summary.pdf. Accessed 29 November 2016.

OECD. 2016b. OECD.Stat Extracts. Available online at http://stats.oecd.org/.

OECD. 2016c. Roadmap to US$100 Billion. Available online at http://dfat.gov.au/international-relations/themes/climate-change/Documents/climate-finance-roadmap-to-us100-billion.pdf.

Olivier, J., T. Leiter, and J. Linke. 2013, November. Adaptation Made to Measure. A Guidebook to the Design and Results-Based Monitoring of Climate Change Adaptation Projects. Deutsche Gesellschaft für Internationale Zusammenarbeit (GIZ) GmbH, Eschborn.

Olsen, G.R. 2001. European Public Opinion and Aid to Africa: Is There a Link? *The Journal of Modern African Studies* 39 (04): 645–674.

Parliament of the United Kingdom. 2015. International Development (Official Development Assistance Target) Act 2015. Chapter 12. Available online at http://www.legislation.gov.uk/ukpga/2015/12/pdfs/ukpga_20150012_en.pdf.

Pickering, J., J. Skovgaard, S. Kim, J.T. Roberts, D. Rossati, M. Stadelmann, and H. Reich. 2015. Acting on Climate Finance Pledges: Inter-Agency Dynamics and Relationships with Aid in Contributor States. *World Development* 68: 149–162.

Raabe, S., B. Kofler, B. Hendricks, S. Rebmann, and K. Roth. 2013. Eine Welt, eine Zukunft: Für eine nachhaltige und fortschrittliche sozialdemokratische Entwicklungspolitik. Arbeitsgruppe für wirtschaftliche Zusammenarbeit und Entwicklung der SPD-Bundestagsfraktion, Berlin.

Richards, L. 2014. *Handling Qualitative Data: A Practical Guide*. London: Sage Publications.

Roodman, D. 2007. The Anarchy of Numbers: Aid, Development, and Cross-country Empirics. Center for Global Development Working Paper Number 32.

Smit, B., and J. Wandel. 2006. Adaptation, Adaptive Capacity and Vulnerability. *Global Environmental Change* 16 (3): 282–292.

SOPAC. 2004. Building Resilience in SIDS: The Environmental Vulnerability Index. Secretariat of the Pacific Community Applied Geoscience and Technology Division (SOPAC). Available online at http://www.sopac.org/index.php/environmental-vulnerability-index.

Teorell, J., S. Dahlberg, S. Holmberg, B. Rothstein, F. Hartmann, and R. Svensson. 2015. The Quality of Government Standard Dataset, version Jan15. University of Gothenburg: The Quality of Government Institute, http://www.qog.pol.gu.se.

Turner, B.L., R.E. Kasperson, P.A. Matsone, J.J. McCarthy, R.W. Corell, L. Christensen, N. Eckley, J.X. Kasperson, A. Luers, M.L. Martello, C. Polsky, A. Pulsiphera, and A. Schiller. 2003. A Framework for Vulnerability Analysis in Sustainability Science. *Proceedings of the National Academy of Sciences* 100 (14): 8074–8079.

UN Statistics Division. 2015. United Nations Commodity Trade Statistics Database. Available online at http://comtrade.un.org/db/.

UNFCCC. 2009. Report of the Conference of the Parties on Its Fifteenth Session, held in Copenhagen from 7 to 19 December 2009. Addendum. Part Two: Action Taken by the Conference of the Parties at Its Fifteenth Session. Document FCCC/CP/2009/11/Add.1.

Voeten, E., A. Strezhnev, and M. Bailey. 2009. United Nations General Assembly Voting Data. Available online at https://dataverse.harvard.edu/dataset.xhtml?persistentId=hdl:1902.1/12379.

Wasielewski Ahlfors, S. 2011. Adapting to a Changing Climate: The Swedish Government's Special Climate Change Initiative 2009–2012. Swedish International Development Agency (Sida), Stockholm.

Weiler, F., C. Betzold, and M. Dornan. 2017. Vulnerability or Donor Interests? The Allocation of Aid for Climate Change Adaptation. Paper presented at the Comparative Politics Seminars, 5 April 2017, University of Kent, Canterbury, UK.

World Bank. 2016. World Development Indicators. Available online at http://databank.worldbank.org/Data/Home.aspx.

Younas, J. 2008. Motivations for Bilateral Aid Allocation: Altruism or Trade Benefits. *European Journal of Political Economy* 24: 661–674.

CHAPTER 4

Descriptive Analysis: Adaptation Aid Flows in the OECD CRS

Before we statistically test the drivers of adaptation aid allocation and qualitatively explore underlying decision-making processes, we want first to get a broad picture of adaptation aid flows. As described in greater detail in Chap. 3, the OECD introduced a so-called Rio marker for adaptation, or adaptation marker, in 2009. This marker allows us to trace adaptation aid flows over time, across donors, and across recipients. Again, however, we want to stress that the adaptation marker and the OECD CRS are not without problems. The data suffer from inconsistent application of the marker, over-reporting, and double-counting of projects towards multiple objectives. These limitations should be kept in mind throughout this and the following chapter.

In this chapter, we provide an overview of adaptation aid from 2010, when the adaptation marker was applied for the first time. We first position bilateral adaptation aid commitments vis-à-vis overall development aid and then position Germany, Sweden, and the UK among all OECD donors. Finally, we also take a look at recipients and examine which countries receive how much adaptation aid, in absolute terms as well as on a per capita basis.

We show that adaptation aid has grown over time, even if it still remains a very small share of overall development aid—about 5% of all development aid reported in the OECD CRS went into adaptation projects. The share of development aid given to adaptation, however, varies across donors. We have already mentioned that Germany, Sweden, and the UK are large

C. Betzold and F. Weiler, *Development Aid and Adaptation to Climate Change in Developing Countries*,
https://doi.org/10.1007/978-3-319-64510-0_4

climate and adaptation donors, which is why we selected them as case studies. As we will see, Germany is particularly important in terms of absolute amounts, while Sweden is most generous on a per capita basis. Other donors, in contrast, commit very little development aid to climate change adaptation.

Our statistical analysis includes bilateral adaptation aid commitments to specific countries only. We thus cover the majority of adaptation aid flows reported in the OECD CRS: most adaptation aid commitments are disbursed rather quickly, and most is given bilaterally. Nonetheless, some donors provide considerable shares of their adaptation aid to unspecified recipients and regional programmes. For instance, Sweden provided only half of its overall adaptation aid to individual countries. From the recipient side, we also see differences: for some recipients, almost all of the support for adaptation came from bilateral donors, while for others, almost all support came from multilateral funds. Recipients also vary with regard to how much adaptation aid they receive. While large countries—unsurprisingly—receive more adaptation aid in total, small island states are among the largest recipients on a per capita basis.

Overall, this chapter provides a broad overview of adaptation aid flows. It shows that our statistical analysis covers a large share of adaptation finance, though we exclude some adaptation aid flows, which may be rather significant for individual donors and/or recipients. The chapter also shows that our selected donors—Germany, Sweden, and the UK—are important adaptation aid donors.

4.1 Overview of Adaptation Aid Flows

4.1.1 Adaptation Aid Compared to Total Aid

How much adaptation aid did developing countries receive since the adaptation marker was introduced? Figure 4.1[1] reports all development aid flows based on the adaptation marker (including ODA loans and grants, equity investments, and other official flows). Remember that the adaptation marker can have four values: the project has not been screened for adaptation objectives; the project has been screened and does not target adaptation; the project has adaptation as a significant objective; or the project has adaptation as its principal objective (see OECD 2011; Chap. 3).

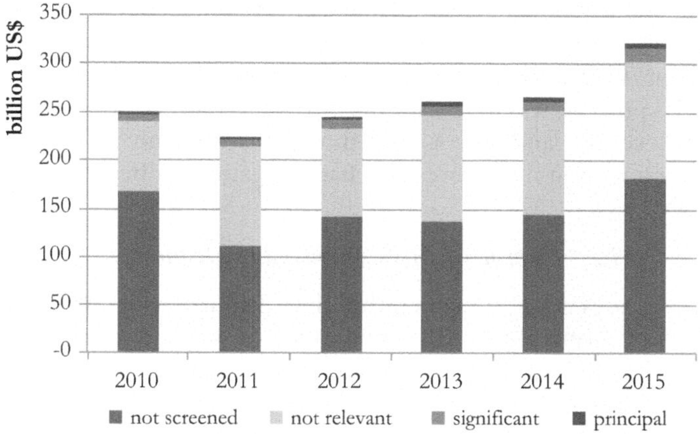

Fig. 4.1 Total aid flows by year and adaptation objective

On average, yearly flows from developed to developing countries reported in the OECD CRS amount to around $260 billion, with an increasing trend since 2011. Aid flows thus reached $323 billion in 2015, not least because more and more donors report their aid to the OECD CRS. Adaptation aid flows are also increasing. While in 2010, donors provided $8.9 billion—3.6% of total development aid—for projects with adaptation as either a significant or the principal objective, this number had increased to $20.3 billion in 2015, or 6.3% of total development aid. Over the 6 years for which we have data, donors thus committed just under $75 billion for adaptation in developing countries, or 4.8% of total development aid.

For about two-thirds of all adaptation-relevant projects worth $50.9 billion, adaptation was a significant objective. In other words, these projects mainly focused on something else, for instance climate change mitigation, but had adaptation co-benefits. Our fictitious project setting up solar panels in Togo but also raising awareness about climate change impacts from Chap. 3 would fall under this category: it is mainly about renewable energy and hence climate change mitigation, but the awareness raising part is (also) about climate change adaptation. The project would thus likely have mitigation as its principal objective and adaptation as a significant objective. Note that we here provide the total amount of funding for such a project, even if only a part of the project (and hence of the funding) was about climate change adaptation.[2] In contrast, adaptation was the principal objective for one-third of all adaptation-relevant projects worth $23.9 billion. These projects would not have taken place without the adaptation component. Our second fictitious project in Togo from Chap. 3 is such a project. Its purpose was to replant mangroves to reduce the risk of flooding from sea-level rise and storms and thus focuses mainly on adaptation—although replanting mangroves also has mitigation co-benefits.

4.1.2 Commitments Compared to Disbursements

Our analysis uses commitments rather than disbursements. Commitments better reflect recent donor decisions, since disbursements partly depend on recipients fulfilling certain conditions (Berthélemy 2006a, 80). Commitments represent 'a firm obligation, expressed in writing and backed by the necessary funds, undertaken by an official donor to provide specified assistance to a recipient country or a multilateral organisation' (OECD 2016b). Although it may take several years, most committed resources are disbursed. Hudson (2013) concludes from his analysis of OECD data from 2002 to 2010 that almost all commitments are met within just 2 years. Yet, the author also finds variation by sector and recipients; for some countries and some sectors, disbursements fall short of commitments. In other words, some donors promise more development aid than they eventually pay out (Hudson 2013).

How accurately does our analysis reflect actual pay-outs? What is the ratio of disbursements to commitments for adaptation aid? Panel (a) of Fig. 4.2 compares committed and disbursed adaptation aid, including both significant and principal adaptation aid. Of course, the yearly commitments and disbursements are not directly comparable, as it may take several years before funds committed in a given year are disbursed. Nonetheless, Fig. 4.2 illustrates trends over time in both committed and disbursed adaptation aid. As expected, disbursements are below commitments throughout the period of analysis. In total, donors have committed $74.7 billion to adaptation since 2010 and disbursed $47.7 billion. Put differently, 70.6% of the resources committed to adaptation between 2010 and 2015 were also disbursed in that same period.

Panel (b) of Fig. 4.2 shows the ratio of commitments to disbursements for selected donors. We selected the three donors with the highest and lowest ratios, our three case study countries Germany, Sweden, and the UK, as well as the OECD average. The figure indicates considerable

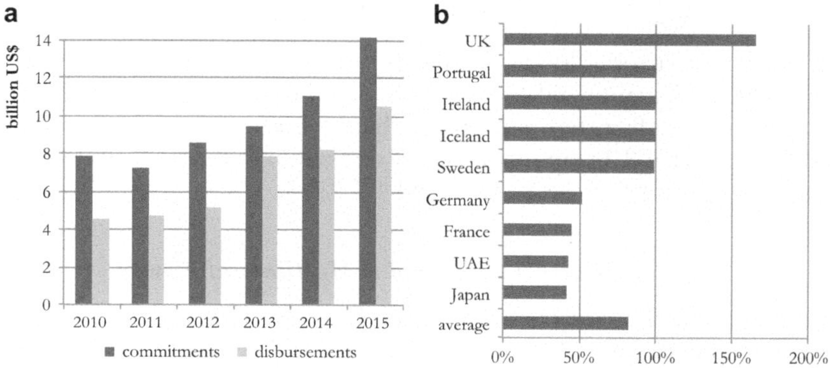

Fig. 4.2 Adaptation aid disbursements compared to commitments over time (**a**); disbursements in percent of commitments for selected donors (**b**)

differences, in line with the analysis by Hudson (2013). The average OECD donor included in the statistical analysis (see Chap. 3) has disbursed 82.3% of the amount it committed. Eleven donors disbursed even 90% or more of their commitments, and 21 disbursed at least two-thirds of their commitments. For the UK, disbursements even exceeded commitments: it committed \$4.4 billion but disbursed \$7.3 billion—a ratio of 165%. This indicates that the UK marked parts of the funding committed prior to 2010, that is, before the Rio marker for adaptation was introduced, as relevant for adaptation but disbursed this funding after 2010. Six donors (Australia, Greece, Iceland, Ireland, Luxembourg, and Portugal) have disbursed all of their committed resources. Sweden disbursed over 99% of what it committed. There are, however, also several donors which committed significantly more adaptation aid than they disbursed. Germany, for example, pledged \$11.9 billion for adaptation, but so far has only paid out about half that amount (\$6.2 billion). Only Korea, France, the United Arab Emirates, and Japan have worse disbursement-to-commitment ratios and disbursed even less than half their committed aid (46.9%, 45.5%, 42.9%, and 41.9%, respectively).

4.2 Who Gives Adaptation Aid?

It may be easier for a donor to fulfil its promises when it has committed only a little adaptation aid to start with. So who gives how much aid for adaptation? In a first step, we compare bilateral to multilateral adaptation aid flows (Sect. 4.2.1), and in a second step zoom in on bilateral donors, the focus of this book (Sects. 4.2.2 and 4.2.3).

4.2.1 Bilateral Compared to Multilateral Flows

Although not all multilateral climate funds report to the OECD CRS, the overall trend is fairly clear: most adaptation aid is provided bilaterally (Fig. 4.3; see also Ayers and Abeysinghe 2013; Weikmans 2016). Of the total \$50.9 billion of committed significant adaptation aid, over 80% (\$41.5 billion) came through bilateral channels, compared to less than 20% (\$9.3 billion) through multilateral channels. Of the total \$23.8 billion of committed principal adaptation aid, three-quarters came from bilateral donors (\$17.8 billion), compared to one-quarter from multilateral donors (\$6.0 billion).

Most multilateral funding is from just one donor, the European Union, which is responsible for 86.6% (\$8.1 billion) of all multilateral significant adaptation aid and for 28.6% (\$1.7 billion) of all multilateral principal adaptation aid reported in the OECD CRS. In total, only a handful of multilateral funds register their adaptation aid in the OECD CRS. Apart from the European Union, the Climate Investment Funds, the European Bank for Reconstruction, the International Fund for Agricultural Development,

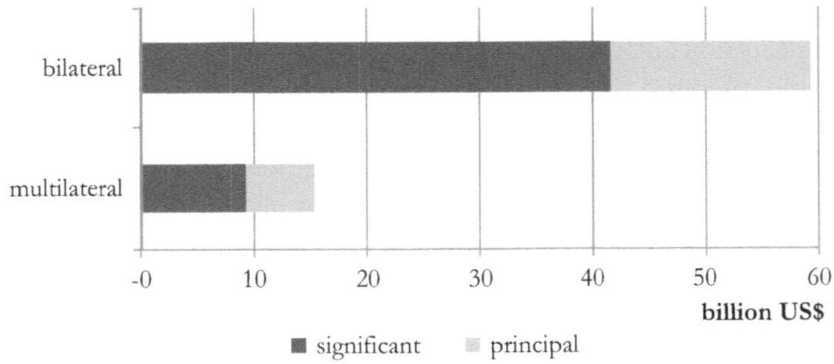

Fig. 4.3 Adaptation aid flows by donor type

the Global Environment Facility, the Adaptation Fund, and the Nordic Development Fund all report adaptation aid flows to the OECD. Even if the Green Climate Fund and other funds such as the Least Developed Country Fund or the Special Climate Change Fund were included, we would still find that the majority of adaptation funding is provided bilaterally. The Climate Funds Update[3] is an alternative source of information on multilateral adaptation (and mitigation) finance. As of September 2016, the database lists a total $33 billion in pledges for multilateral adaptation funds—this is still just about half of the $58 billion in bilateral pledges reported in the OECD CRS (see also Ayers and Abeysinghe 2013; Weikmans 2016).

That donors prefer bilateral over multilateral channels is well documented and has to do with donor influence. While donor countries can in principle freely decide which countries and projects to support with their bilateral aid, they have only limited influence over multilateral aid flows (see e.g. Gulrajani 2016). This is an additional reason for us to analyse bilateral rather than multilateral adaptation aid.

4.2.2 *Bilateral Donors in Comparison: Total Adaptation Aid*

Let us take a closer look at the 28 bilateral donors included in the statistical analysis of the next chapter. Figure 4.4 lists adaptation aid flows for selected donors, including again the least and most generous donors, our three case study countries, as well as the average for all bilateral donors.

The average OECD donor committed $1.5 billion to significant and $640 million to principal adaptation projects, or a total of $2.1 billion for adaptation to climate change, but there are stark differences between individual donors in how much aid they provide for the latter. Partly, these differences stem from the different overall aid budgets. Some countries provide much more development aid in general, as we would expect. After all, the donor countries differ strongly in terms of their population and/or GDP. It is but natural that Iceland, with a population of 300,000, spends less on development aid than the USA with a population of over 300 million. Countries that provide less aid overall may also provide less for climate change purposes. Additionally, donors have different focal areas; some may thus focus their aid programmes in areas such as education or health rather than climate change—although climate change of course affects aid projects in all areas and should always be taken into account (see Chap. 2).

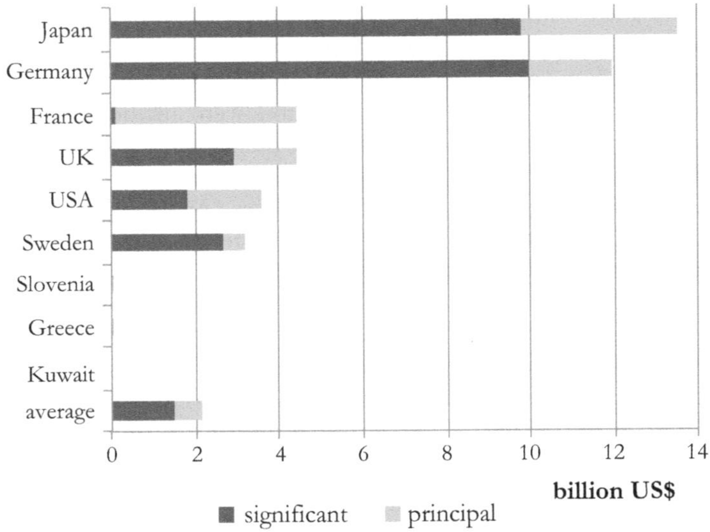

Fig. 4.4 Total adaptation aid for selected donors

The largest adaptation donors in absolute terms are also large overall donors (Fig. 4.4). Japan and Germany provided \$13.5 and \$11.9 billion, respectively, for adaptation projects between 2010 and 2015. Mostly, these countries supported projects with a significant adaptation objective, with \$10.0 and \$9.8 billion, respectively, going to such projects, compared to \$3.7 and \$2.0 billion going to projects with adaptation as their principal objective. In other words, 72.5% of Japanese and 83.6% of German adaptation aid supported projects where adaptation was a significant objective. This focus on significant adaptation aid was the case for almost all donors; 19 of the 28 included in our statistical analysis gave more than two-thirds to significant adaptation projects, and 16 of the 28, more than three-quarters. France, the third-largest adaptation donor, is the exception to the rule. Almost all French adaptation went to principal adaptation projects: of the \$4.4 billion, \$4.3 billion (96.9%) supported such projects, and only \$135 million (3.1%) supported significant adaptation projects. Just like France, the UK provided \$4.4 billion for adaptation projects, but it split this funding more like the average donor: two-thirds of British adaptation aid (\$2.9 billion) was for projects with significant adaptation objectives, and one-third ($1.5 billion) for projects with adaptation as the principal objective. The USA falls somewhere in between. It committed $3.6 billion to adaptation and split this almost evenly between significant and principal adaptation funding. Sweden, our third case study country, is also a fairly large adaptation donor in absolute terms—although it is much smaller than Japan, France, Germany, the UK, and the USA. Sweden committed $3.2 billion for climate change adaptation and spent this funding mainly on significant adaptation projects ($2.6 billion or 82.6%) rather than on principal adaptation projects ($558 million or 17.4%). Partly because of their small size, Slovenia, Greece, and Kuwait provided the least adaptation aid in absolute terms. Slovenia spent most of its $7.3 billion adaptation aid on significant adaptation aid projects ($6.7 billion or 91.3%), while Greece spent most of its $5.8 billion adaptation aid on principal adaptation projects ($4.9 billion or 84.5%). Kuwait did not mark any of its development aid as relevant for adaptation.

4.2.3 Bilateral Donors in Comparison: Per Capita and Percent

Comparing total adaptation aid flows across donors may be misleading, however, because of the different sizes and different aid budgets of the different donor countries. We have already mentioned that the USA, for instance, has a population of over 300 million, while the population of Iceland is only 330,000—about 0.1% of the US population (figures from United Nations Statistics Division 2014). Panel (a) of Fig. 4.5 thus compares selected OECD donors with regard to their adaptation aid on a per capita basis, that is, how much each inhabitant of the donor country paid for adaptation aid to developing countries between 2010 and 2015. Panel (b) compares selected donors with regard to how much in percent of their overall development aid was spent on climate change adaptation.

On a per capita level, the Scandinavian countries were by far the most generous adaptation donors. Each inhabitant of Sweden thus gave about $330 for adaptation projects in the Global South, mostly for projects with significant adaptation objectives. Danes and Norwegians gave comparable amounts, at $327 and $299 per capita, respectively, again mainly for significant adaptation projects. Germans gave about half that sum: $148 per capita, still far above the OECD average. The UK committed $68 per capita to adaptation, much closer to the overall OECD average. If the total adaptation aid of $59.4 billion provided by the 28 donor

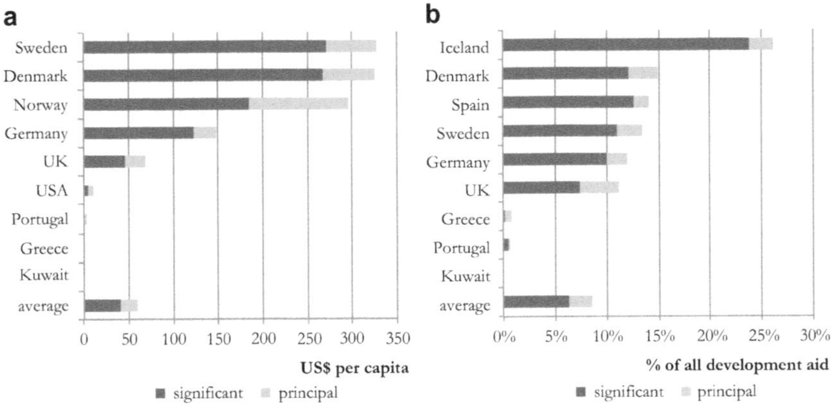

Fig. 4.5 Adaptation aid for selected donors; per capita (**a**), and in percent of all development aid (**b**)

countries considered in our statistical analysis[4] were split evenly among their inhabitants, each person would have given $59 for adaptation: $41 for significant adaptation projects and $18 for principal adaptation projects. In practice, many OECD donors gave less on a per capita basis. In the USA, per capita adaptation aid amounted to $11; only Italy ($8), Slovenia ($3.50), the Czech Republic ($1.50), Portugal ($1.30), Greece ($0.50), and Kuwait (no adaptation aid) committed less per capita.

When we look at how much in percent of a donor's total development aid budget was marked as relevant for adaptation, we get a slightly different picture. Scandinavian countries again rank highest. While the Icelandic adaptation aid budget overall is rather small because of the country's small size, it dedicated over one-quarter (26.1%) of its development aid to adaptation projects. (Icelandic adaptation aid was also above average on a per capita basis, at $107.) Denmark, Finland, and Sweden also dedicated relatively much of their development aid to adaptation, with 14.8% of Danish, 13.9% of Finnish, and 13.4% of Swedish aid having adaptation objectives. Norway is the exception here, with only 5.8% of its aid targeting adaptation. Spain (14.2%), Germany (12.0%), and the UK (11.2%) dedicated similar shares of their development aid to adaptation as did the Scandinavian countries. In contrast, less than 1% of Greek and Portuguese development aid flowed into adaptation projects. Kuwait, as mentioned earlier, did not mark any of its development aid as relevant for adaptation. Overall, 7.6% of total development aid addressed adaptation. Mostly, adaptation was a significant objective (6.3% of total development aid) rather than the principal objective (2.3% of total development aid).

The comparison of adaptation across donors—whether with regard to total adaptation aid flows, per capita adaptation aid, or adaptation as a percentage of total development aid—reveals stark differences. Overall, Scandinavian countries gave fairly high levels of adaptation aid, notably when we consider per capita adaptation aid and adaptation aid as a percentage. This mirrors general trends in development aid; the development aid literature describes Scandinavian countries as most generous (e.g. Berthélemy 2006b). Our three case study countries also are rather consistently among the largest adaptation donors. In contrast, Southern European countries like Portugal or Greece give relatively little adaptation aid. Kuwait is the only donor country in the OECD CRS that did not mark any of its development as targeting adaptation.

4.3 Who Receives Adaptation Aid?

Having compared adaptation aid donors in the last section (Sect. 4.2), we now compare adaptation aid recipients. We first look at the type of recipients, since donors provide adaptation aid not only to individual countries, but also to regional programmes or unspecified recipients (Sect. 4.3.1). While most adaptation aid in the period 2010–2015 targeted individual countries, some donors, for instance Sweden, allocated fairly high shares of their adaptation aid to regional programmes or unspecified recipients—these flows are not included in our statistical analysis. We then look at bilateral compared to multilateral adaptation aid flows (Sect. 4.3.2). As we have seen in Sect. 4.2.1, most adaptation aid was given bilaterally, but for some recipients, almost all support came from multilateral funds. Finally, we turn to the distribution of bilateral adaptation aid across individual countries. Here, we examine total adaptation aid flows as well as adaptation aid per capita (Sect. 4.3.3). We show that the level of adaptation aid that countries receive varies strongly. Pacific island countries, for instance, received very little in absolute terms, but very high levels per capita.

4.3.1 Type of Recipient

Our analysis focuses on the distribution of adaptation across individual recipient countries. Yet donors can also support regional programmes. Similarly, domestic development related costs—that is, resources spent *within* the donor country—count as ODA. Costs related to refugees within the donor country, administrative costs, or research costs are reported to the CRS as development aid. In this case, the recipient is unspecified (OECD 2016a). This is also the case if a specific project benefits more than one region (OECD 2016d).

Which part then of bilateral adaptation aid does our analysis cover? How much adaptation aid is targeted at individual countries? How much is delivered via regional programmes? And how much stays with the donor or is given to inter-regional programmes? Figure 4.6 shows the total adaptation aid flows according to these three types of recipients. Panel (a) plots the distribution over recipient types across all donors, while panel (b) displays the distribution for selected donors.

Overall, by far the largest share of adaptation aid –$42.3 billion, or 71.4%—was given bilaterally, that is, to individual countries. $5.9 billion (10.0%) was given via regional programmes, and $11.1 billion (18.7%) had unspecified recipients. The distribution is fairly similar when looking at significant and principal adaptation aid separately. Just over 70% of

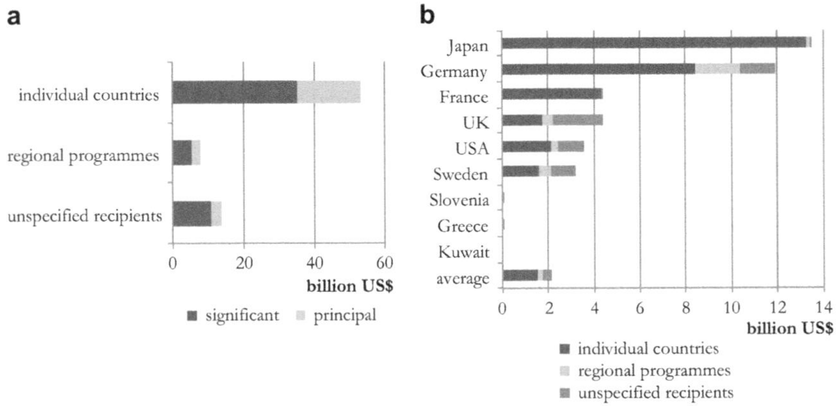

Fig. 4.6 Adaptation aid by recipient type; overall (**a**), and for selected donors (**b**)

significant as well as principal adaptation aid is directly targeted at individual countries. Slightly more significant adaptation aid went to unspecified recipients (20.1%) compared to principal adaptation aid (15.5%).

However, there are large differences across donors in how they allocate their adaptation aid to different recipient types, as panel (b) of Fig. 4.6 indicates. While donors overall clearly prefer to provide most of their adaptation aid bilaterally to individual recipient countries, some donors instead focus most of their adaptation aid on regional programmes or unspecified recipients. The United Arab Emirates, France, Luxembourg, and Japan provided all or almost all of their adaptation aid bilaterally to individual countries (100%, 98.7%, 98.4%, and 98.1%, respectively). Germany also had a clear preference for supporting individual countries at 70.6%. In contrast, three smaller countries—the Netherlands, Greece, and Iceland—provided less than one-third of their adaptation aid to individual countries (31.1%, 29.5%, and 28.0%, respectively). Instead, these countries did not specify any recipient for much of their adaptation aid at 59.3% for the Netherlands, 47.3% for Greece, and 59.1% for Iceland. Similarly, 49.9% of British, 32.5% of Swedish, and 12.8% of German adaptation aid had no specified recipient.

Donors spent relatively small shares of their adaptation aid on regional programmes. Of the 28 donor countries, 12 allocated less than 5% of their adaptation aid budget to regional programmes. In contrast, four donors—Greece, Finland, Austria, and Canada—gave over 20% of their adaptation aid budget to regional programmes. Austria distributed 36.8%, Canada half its adaptation aid on regional programmes. Sweden and Germany also gave comparatively a lot to regional programmes (18.0% of Swedish and 16.6% of German adaptation aid). The UK spent 9.6% of its adaptation aid on regional programmes, very close to the OECD average of 10%.

Note that our statistical analysis in the next chapter excludes all funds to regional programmes as well as to unspecified recipients, as do the next sections that compare individual recipient countries.

4.3.2 Bilateral Versus Multilateral Flows

Let us now focus on adaptation aid that goes to individual countries. Which countries receive support for adaptation and how much? Here, we need to distinguish further between multilateral and bilateral adaptation aid. Even if, overall, the majority of adaptation aid was from bilateral donors (see Sect. 4.2.1), some recipient countries received a considerable

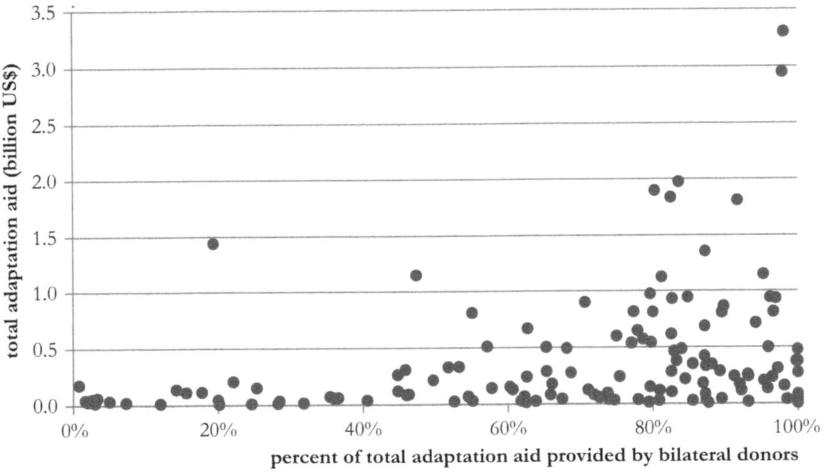

Fig. 4.7 Share of adaptation aid provided by bilateral donors

share, if not most, of their total adaptation aid from multilateral donors. Figure 4.7 plots the total amount of adaptation aid—including projects with significant and principal adaptation objectives—against the share of this aid provided by bilateral donors.

Figure 4.7 indicates two trends. First, the total amount of adaptation aid per recipient rarely exceeded $1 billion, with most countries receiving $500 million or less. Yet, for some countries, support for adaptation was considerably higher than the average of $369 billion. Vietnam obtained most, with $3.3 billion, followed by India, which received $2.9 billion, and Ethiopia, which received $2.0 billion.

Second, the share of total adaptation aid provided by bilateral donors ranges from 0% to almost 100%. Some countries received almost all of their adaptation aid from multilateral donors, and others received all of it from bilateral donors. Overall, however, the majority of countries received more support from bilateral donors compared to multilateral donors. More than three-quarters of the recipients (109 out of 141 recipient countries) received at least 50% of their adaptation aid from bilateral sources, and over half (78 out of 141 recipient countries) received at least 75% from bilateral sources.

Figure 4.7 also suggests a positive, if weak, relationship between the total level of adaptation aid and the share provided by bilateral donors: the higher the total amount of adaptation aid, the higher the share provided by bilateral donors. Among the 11 largest recipients—whose total adaptation aid exceeded $1 billion—only two obtained considerable support from multilateral donors: Turkey received 80.6% of its total $1.4 billion from multilateral sources, and Morocco 52.6% of its total $1.1 billion. For Vietnam and India, the two largest recipients of adaptation aid, more than 95% of their adaptation aid came from bilateral donors (98.2% and 98.0%, respectively). For Ethiopia, the third largest recipient, the share of bilateral adaptation aid was 83.6%. Several smaller countries also received all or almost all of their adaptation aid from bilateral sources. For 28 countries, the share of bilateral adaptation aid exceeded 95% and 13 countries even obtained all of their aid from bilateral donors, including four SIDS.[5] In contrast, seven recipient countries obtained 95% or more of their adaptation aid from multilateral sources, five of which are SIDS.[6]

4.3.3 Recipients in Comparison

Let us now focus on bilateral adaptation aid and exclude multilateral adaptation aid, as our regression analysis in the next chapter does, too. Who received bilateral adaptation aid, and how much? To compare recipient countries, we used two measures of adaptation aid: total adaptation aid flows and adaptation aid per capita. Figure 4.8 shows these measures of adaptation aid for principal adaptation aid, and Fig. 4.9 for significant adaptation aid.

Where does adaptation aid flow? In terms of total *principal* adaptation aid (panel (a) of Fig. 4.8), the largest recipients were rather populous countries in Asia that are—at least at first glance—also fairly vulnerable to climate change: Vietnam ($1.3 billion), the Philippines ($888 million), Bangladesh ($765 million), and Indonesia ($668 million). Other large recipients of principal adaptation aid include Colombia ($656 million), Jordan ($539 million), and Kenya ($397 million). The picture changes only slightly when we consider *significant* adaptation aid (panel (a) of Fig. 4.9): India was the largest recipient with $2.6 billion. Vietnam ($1.9 billion), Ethiopia ($1.3 billion), Kenya ($1.1 billion), the Philippines ($770 million), and Bangladesh ($756 million) remain in the top 10 recipients of adaptation aid. Other countries that received high levels of support for significant adaptation (but not so much for principal adaptation) are Ukraine ($1.1 billion) and Tanzania ($833 million).

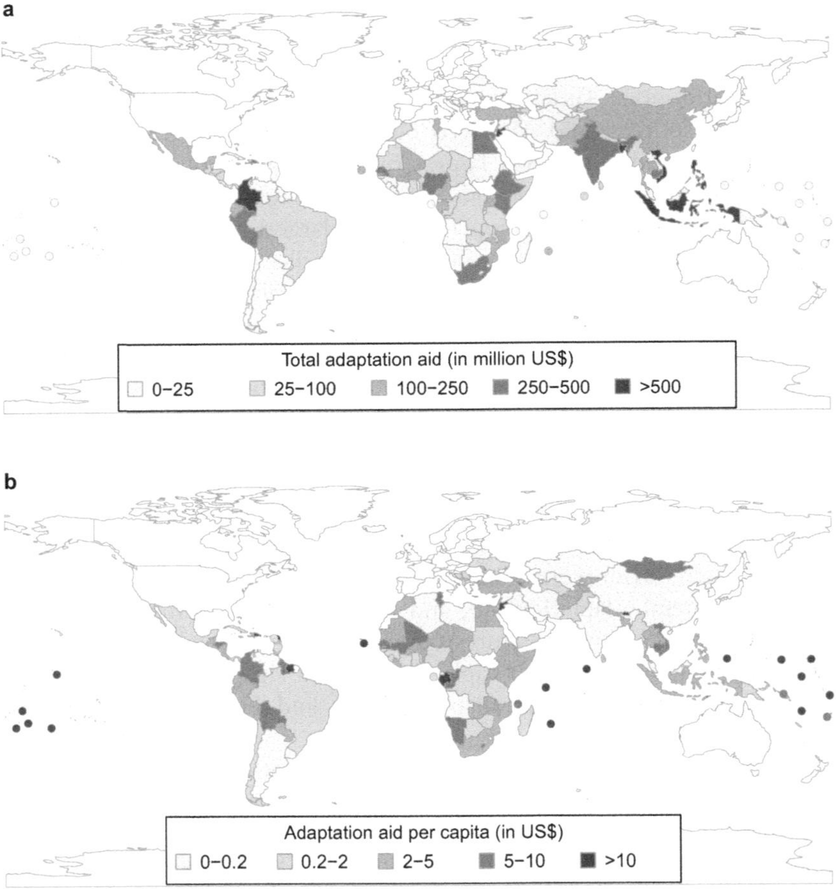

Fig. 4.8 Distribution of principal adaptation aid

Yet comparing absolute amounts of adaptation aid may be misleading; India has a population of over 1 billion, Vietnam has a population of over 90 million. We would expect that they obtain more aid than for example Jordan, which has a population of just under 10 million. Per capita adaptation aid is thus a better measure to compare recipients (panel b of Figs. 4.8 and 4.9). If all bilateral adaptation aid were distributed equally across the population of the Global South, each inhabitant would have

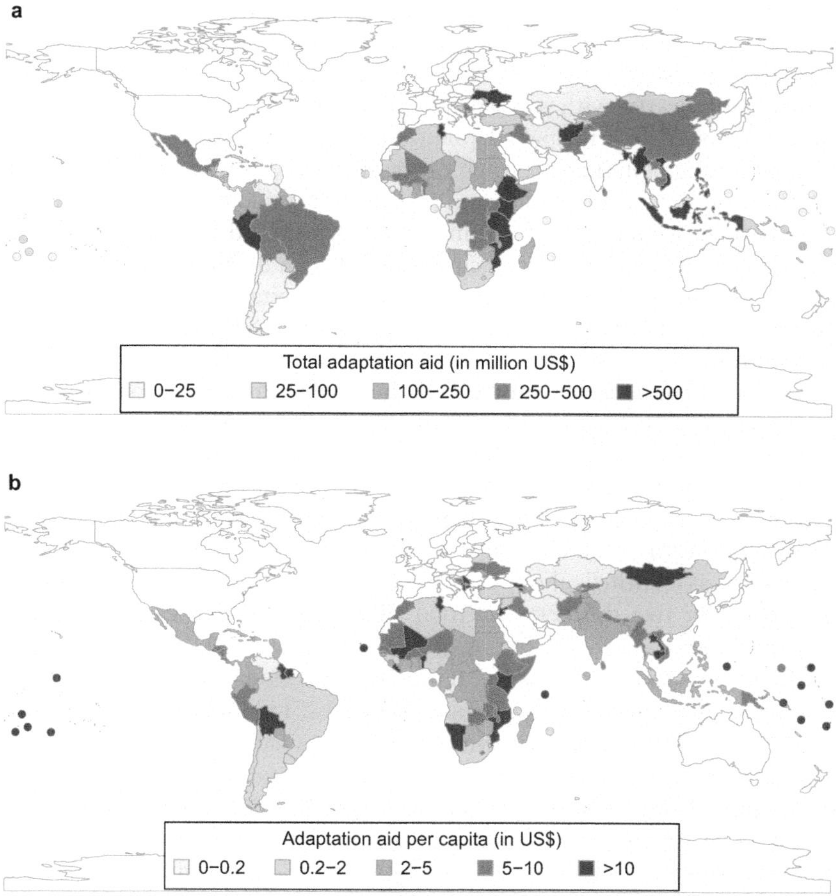

Fig. 4.9 Distribution of significant adaptation aid

received $2.25 for principal and $4.95 for significant adaptation projects. In practice, adaptation aid is distributed very unequally, ranging from just 9 cents in Kazakhstan (1 cent for principal adaptation and 8 cents for significant adaptation) to almost $20,000 in Niue ($2009 for principal adaptation and $17,620 for significant adaptation).[7]

At first glance, differences in population size seem to be able to account for these differences in per capita adaptation aid. The largest recipients of it per capita have the smallest population sizes. The top 10 recipients were almost entirely SIDS with very small populations. Per capita, Tuvalu was the largest recipient of principal adaptation aid and the third-largest recipient of significant adaptation aid: each of the mere 10,000 Tuvaluans received $2486 for principal adaptation and $1115 for significant adaptation. Niue was the largest recipient of significant adaptation aid per capita and the second-largest recipient of principal adaptation aid. Niue's population is only 1500, hence the very high level of per capita adaptation aid: each Niuean received $2009 for principal adaptation and $17,620 for significant adaptation.

Gabon is the largest country and the only non-SIDS among the top 10 per capita adaptation aid recipients. Each of the 1.6 million Gabonese received $135 for principal adaptation projects—but only $2.50 for significant adaptation projects. Largely, the high level of per capita support in Gabon is due to one single adaptation project: France pledged almost $145 million for a sewage and drainage project in Libreville, which had adaptation as its principal objective;[8] $145 million corresponds to $88.80 per capita, or two-thirds of the total adaptation aid that each Gabonese citizen obtained.

Mauritius, also a SIDS, is the tenth-largest recipient of principal adaptation aid per capita; it has a population of 1.2 million. Guyana, another SIDS, is the ninth-largest recipient of significant adaptation per capita; it has a population of almost 800,000. The remaining countries in the top 10 list of per capita recipients all have populations of less than about 500,000.

The large recipients of absolute adaptation aid flows, in contrast, received comparatively little on a per capita basis: each Indian for instance received 20 cents for principal and $2.10 for significant adaptation; each Vietnamese received $15 for principal and $21 for significant adaptation; and each Ethiopian, $3.50 for principal and $14 for significant adaptation. Note though that some fairly small countries also received rather little aid for adaptation per capita (from bilateral donors): Swaziland, with a population of just over 1.2 million, received $1.59 per capita; Equatorial Guinea, with a population of less than 800,000, received $1.77 per capita, all for significant adaptation projects. Belize and St Vincent and the Grenadines, two SIDS, also received rather little: each of the around 340,000 inhabitants of Belize and each of the around 110,000 inhabitants of St Vincent and the Grenadines benefitted from $4.35 and $5.40, respectively, for adaptation.[9]

In this chapter we have sought to give a very broad overview of adaptation aid flows as reported in the OECD CRS, which serves as a backdrop for the statistical analysis in the next chapter as well as the qualitative case studies in Chap. 6. While actual numbers should be treated with caution given the inherent weaknesses of aid data and donors' self-reporting (see Chap. 3), this chapter has indicated some broad trends. First, we started with the observation that adaptation aid remains a small—if growing—share of total development aid, and that adaptation aid is mainly provided bilaterally. Second, when we examine more closely which bilateral donors provide how much adaptation aid, we recognise large differences across donors. Regardless of how we compare adaptation aid flows across donors—total amounts, per capita, or in percent of all development aid—our case study countries of Germany, Sweden, and the UK are among the most important adaptation donors. Finally, we have also compared recipients with regard to how much adaptation aid they obtain. While donors prefer to provide their aid bilaterally to specific countries, they also support regional programmes and report in-country development-related expenses. When we look at individual countries, large populous countries received most adaptation aid in absolute terms, while very small countries like some SIDS received most adaptation aid per capita. In the next chapter, we will test more systematically how donors allocated their bilateral adaptation aid, and what role size and other factors play.

Notes

1. All figures in this chapter are based on data from OECD (2016c) and are in constant 2013 US$.
2. Since projects with adaptation as a significant objective are only partly about climate change adaptation, we discount significant adaptation aid at 50% in our statistical analysis. This is in line with how many donors report their significant adaptation (and mitigation) aid as climate finance (see e.g. Adaptation Watch 2015, 32).
3. The Climate Funds website is a joint initiative of the Heinrich Böll Foundation and the Overseas Development Institute. See http://www.climatefundsupdate.org/ (accessed 7 February 2017).
4. Together, these 28 donors account for over 99% of all bilateral adaptation aid in the OECD CRS.
5. These 13 countries are Cabo Verde, Cameroon, Congo, Equatorial Guinea, Fiji, Iran, Libya, Marshall Islands, Micronesia (Federated States), North Korea, Panama, Syria, and Togo.

6. These seven countries are Dominica, Kazakhstan, St Kitts and Nevis, St Lucia, St Vincent and the Grenadines, São Tomé and Príncipe, and Swaziland.
7. Like the Cook Islands, Niue is freely associated with New Zealand and traditionally receives very high levels of support from the latter. It has even been described as 'the world's most aid-dependent country' (Barnett 2008, 33).
8. See http://gabon.afd.fr for more information on the project 'Aménager le bassin versant de Gué Gué'. Single projects have a similar impact in other small countries (see Betzold 2016).
9. St Vincent and the Grenadines did, however, receive considerable support for adaptation from multilateral funds.

References

Adaptation Watch. 2015. Toward Mutual Accountability: The 2015 Adaptation Finance Transparency Gap Report. Policy Briefing. Available online at http://www.adaptationwatch.org/.

Ayers, J.M., and A.C. Abeysinghe. 2013. International Aid and Adaptation to Climate Change. In *The Handbook of Global Climate and Environmental Policy*, chap. 28, ed. R. Falkner, 486–506. New York: Wiley.

Barnett, J. 2008. The Effect of Aid on Capacity to Adapt to Climate Change: Insights from Niue. *Political Science* 60(1): 31–45.

Berthélemy, J.-C. 2006a. Aid Allocation: Comparing Donors' Behaviours. *Swedish Economic Policy Review* 13: 75–109.

Berthélemy, J.-C. 2006b. Bilateral Donors' Interest vs. Recipients' Development Motives in Aid Allocation: Do All Donors Behave the Same? *Review of Development Economics* 10(2): 179–194.

Betzold, C. 2016, Aid and Adaptation to Climate Change in Pacific Island Countries. Development Policy Centre Discussion Paper 46. Development Policy Centre, Australian National University, Canberra.

Gulrajani, N. 2016. Bilateral versus Multilateral Aid Channels: Strategic Choices for Donors. ODI Report. Available online at https://www.odi.org/sites/odi.org.uk/files/resource-documents/10492.pdf, Overseas Development Institute, London.

Hudson, J. 2013. Promises Kept, Promises Broken? The Relationship Between Aid Commitments and Disbursements. *Review of Development Finance* 3: 109–120.

OECD. 2011. Handbook on the OECD-DAC Climate Markers. Available online at www.oecd.org/dac/stats/48785310.pdf.

OECD. 2016a. Country Programmable Aid (CPA): Frequently Asked Questions. Available online at http://www.oecd.org/development/effectiveness/countryprogrammableaidcpafrequentlyaskedquestions.htm.

OECD. 2016b. DAC Glossary of Key Terms and Concepts. Available online at http://www.oecd.org/dac/dac-glossary.htm. Accessed 9 November 2016.

OECD. 2016c. OECD.Stat Extracts. Available online at http://stats.oecd.org/.

OECD. 2016d. Technical Guide to terms and data in the Creditor Reporting System (CRS) Aid Activities database. Available online at http://www.oecd.org/dac/stats/crsguide.htm.

United Nations Statistics Division. 2014, December. Exchange Rates and Population. National Accounts Main Aggregates Database. Available online at http://unstats.un.org/unsd/snaama/dnlList.asp.

Weikmans, R. 2016. Dimensions éthiques de l'allocation du financement international de l'adaptation au changement climatique. *VertigO - la revue électronique en sciences de l'environnement* 16(2).

CHAPTER 5

Quantitative Analysis: Who Gives Adaptation Aid to Whom?

Chapter 4 suggested that size can partly explain why some countries receive a lot of support for adaptation to climate change and others very little. Vulnerability, at least at first glance, also seemed to matter. In this chapter, we analyse the distribution of adaptation aid in more detail and more systematically. Applying multivariate regression analysis techniques, we test our hypotheses regarding recipient need, recipient merit, and donor interests (see Chap. 2). We first want to understand what role these three factors play across all donors and then examine how our three case study countries—Germany, Sweden, and the UK—allocate their adaptation aid.

Our regression analysis makes use of the dyadic dataset that we described in detail in Chap. 3. This dataset records how much adaptation aid per capita (in constant 2013 US$) each donor committed to each recipient in every year covered. We distinguish between two dependent variables: one considers principal adaptation aid only, the other also includes significant adaptation aid, discounted at 50%. To evaluate how donors allocate their adaptation aid, we estimate two stages of the decision-making process (see Chap. 3): at the selection stage, we examine whether or not recipient countries receive any adaptation aid at all, regardless of the amount. In this case, our dependent variable is a dummy variable that takes a value of 1 if recipient i receives any adaptation aid from donor j in year t. We estimate this selection stage with logit models with donor random effects and year fixed effects. At the allocation stage, we examine how much adaptation

C. Betzold and F. Weiler, *Development Aid and Adaptation to Climate Change in Developing Countries*,
https://doi.org/10.1007/978-3-319-64510-0_5

aid recipient countries receive, but only consider those countries that did obtain some adaptation aid at the first stage. Our dependent variable at this stage is the amount of adaptation aid recipient country *i* receives from donor *j* in year *t*. We estimate this allocation stage with linear models, again with donor random effects and year fixed effects.

We first present the results of the analysis of a pooled sample of all donor countries that report aid data to the OECD CRS. In a second step, we separately run the same models for our three selected case study countries (Germany, Sweden, and the UK). We are thus able to identify clearly which factors determine adaptation aid allocation in general, which explanations are only valid in some countries and under specific circumstances, and which factors do not play a role at all.

Overall, we find that vulnerability matters, at least its physical dimension: countries that are more exposed and sensitive to climate risks tend to receive more adaptation aid, all else being equal. The findings are more ambiguous with regard to adaptive capacity, which is closely linked to good governance and thus recipient merit, and it is therefore controversial to what extent countries with low adaptive capacity are themselves responsible for their higher level of vulnerability—and to what extent donors would thus 'reward' poor governance by allocating more adaptation aid to these countries (see Füssel 2009, 18). Our results suggest that donors are more concerned with good governance than with adaptive capacity, as they tend to provide more adaptation aid to well-governed countries—although poverty is a very strong predictor of who receives adaptation aid, and how much. Further, we find that donors take into account their own interests and notably provide more adaptation aid to countries to which they export a lot. In contrast to other studies, however, the results for our other measures of donor interests are more ambiguous. Finally, our analysis highlights that adaptation aid is closely linked to development aid in general; adaptation aid is a subset of development aid and donors tend to support adaptation in countries in which they also engage in other development projects. And, as Chap. 4 has already suggested, we find evidence for the so-called 'small country bias': the smaller a country—in terms of its population—the more adaptation aid it receives per capita.

When we examine recipient need, recipient merit, and donor interests for Germany, Sweden, and the UK separately, we find that there are considerable differences between individual donors. Not least because of the smaller number of observations, the results for individual donors are generally much weaker than the results for the pooled sample of all donors. Physical vulnerability is a factor for the UK, but not so much for Germany and Sweden. In contrast, Germany and particularly Sweden pay attention to governance in recipient countries. Donor interests are generally of little interest in all three countries. Just as for the pooled sample, we find a close link between total development aid and adaptation aid—a finding we come back to in the qualitative analysis—as well as the 'small country bias'.

5.1 Adaptation Aid Allocation Across All Donors

Before we start modelling adaptation aid allocation using multiple regression techniques, let us first take a look at the correlation coefficients between all independent and dependent variables used in this study. Figure 5.1 displays these correlations graphically. Larger dots represent a stronger correlation between the two variables in question. The grey scale indicates whether the relationship is positive or negative: darker dots represent positive correlations and lighter dots, negative correlations.

The figure first shows that there is a strong positive (and significant) relationship between the two dependent variables used in this study, principal adaptation aid per capita and principal plus discounted significant adaptation aid per capita. This clear relationship between the two response variables is of course to be expected, as the former is a substantial element of the latter. Because we use the two specifications of our dependent variable separately, this does not pose a problem from a methodological point of view. In contrast, the correlation between the two dependent variables and the majority of the independent variables is not particularly strong, as the first and second column/first and second row of Fig. 5.1 show. To some extent, the correlation coefficient between per capita adaptation aid (principal only, as well as principal and significant) and per capita development aid is the exception. This correlation is the highest between the dependent variables and any independent variable, with a correlation coefficient of 0.18 for principal adaptation aid and total development aid, and a correlation coefficient of 0.27 between principal and significant adaptation aid and total development aid. This is a first indication that donors' allocation of *adaptation* aid is strongly based on their allocation of *development* aid, with other factors explaining less (if anything) of these allocation decisions. Our qualitative analysis confirms that these two allocation decisions are not separate decisions (see Chap. 6). We will address the implications of this point in the concluding chapter.

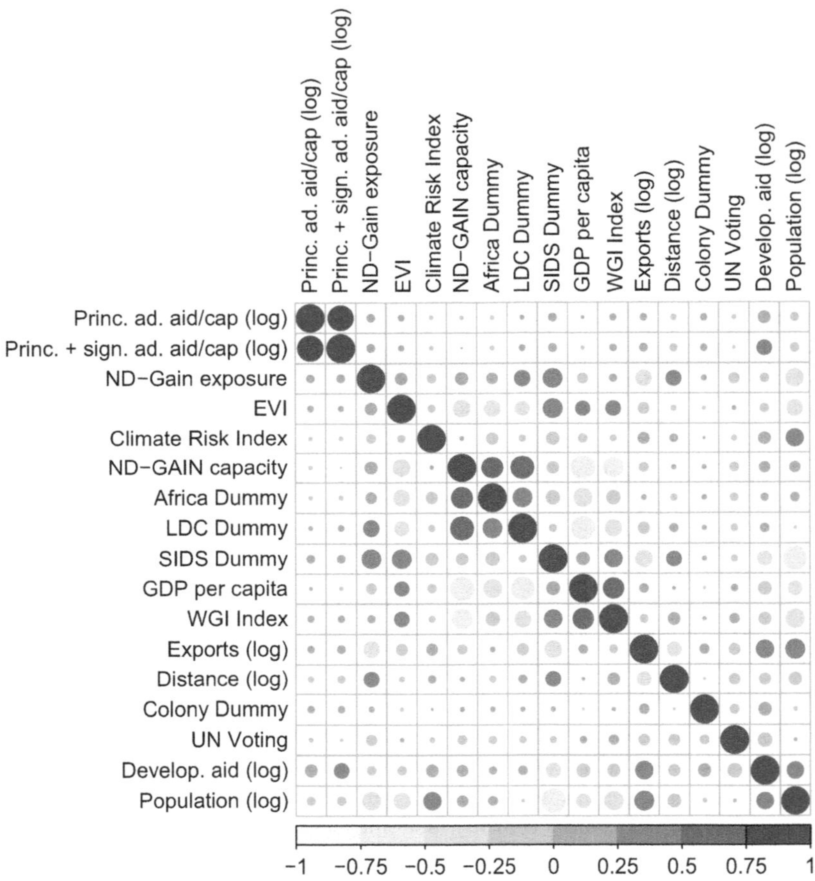

Fig. 5.1 Correlation matrix of all dependent and independent variables

For the independent variables, most of the correlation coefficients seem unproblematic. Unsurprisingly, GDP per capita is quite strongly and negatively correlated with two of our three dummy variables, capturing particularly vulnerable country status, namely African countries and LDCs. Furthermore, GDP per capita is also strongly and negatively correlated with the ND-GAIN adaptive capacity sub-index, with a correlation coefficient of −0.74. The latter is comprised of variables such as percentage of paved roads, child malnutrition, and access to electricity. Higher values on the ND-GAIN adaptive capacity sub-index indicates a lower capacity to deal with the challenges of climate change, and therefore higher vulnerability. Although GDP per capita itself is not included in this sub-index, it strongly influences the other components of the ND-GAIN adaptive capacity sub-index. It is therefore not surprising that we find a strong relationship between GDP per capita and the ND-GAIN adaptive capacity sub-index. Accordingly, the ND-GAIN adaptive capacity sub-index also correlates strongly with two vulnerable country dummies—the dummy for African countries (correlation coefficient of 0.56) and that for LDCs (correlation coefficient of 0.67)—as well as with our composite WGI index (correlation coefficient of −0.52). These correlations go back to the discussion in Chap. 3 about the difficulty of distinguishing between adaptive capacity (H1b) and recipient merit (H2), a difficulty which needs to be considered when interpreting the models presented later. The only other correlation coefficient that is relatively large is the relationship between the SIDS dummy and population (correlation coefficient of −0.72), which should be expected and which is not problematic.

Tables 5.1, 5.2, 5.3, and 5.4 show all the regression results for the two dependent variables and the full set of donor countries. Tables 5.1 and 5.2 display the results of the selection and allocation stage, respectively, for principal adaptation aid, Tables 5.3 and 5.4 list the results of the two stages for principal and discounted significant adaptation aid. In each case, we run four models: the first model includes only our three measures of physical vulnerability to climate change; the second model focuses on adaptive capacity and recipient merit; the third model examines donors interests; and the fourth model is a full model that includes all factors we use to explain adaptation aid allocation. Each model also includes the two control variables: total development aid and population size. Overall, the models for the two different dependent variables are remarkably similar to each other. Instead of discussing the models for the different dependent variables separately, we therefore split the discussion into three parts. First, we discuss the effect of physical exposure and sensitivity to climate change on adaptation aid allocation (H1a). Second, we take a closer look at the impact of adaptive capacity and recipient merit on aid allocation. As adaptive capacity cannot be clearly separated from recipient merit (see the discussion in Chaps. 2 and 3), we examine the variables falling into these two categories together and thus discuss H1b and H2 concurrently. Finally, we turn our attention to the results for donor interest (H3).

Table 5.1 Principal adaptation aid, selection stage, all countries

	Dependent variable: principal adaptation aid—Yes/No			
	(1)	(2)	(3)	(4)
ND-GAIN exposure	3.193***			2.917***
	(0.473)			(0.567)
EVI	−0.045			−0.049
	(0.039)			(0.048)
CRI	0.009***			0.006***
	(0.001)			(0.001)
African dummy		−0.423***		−0.482***
		(0.069)		(0.077)
LDCs		0.104		0.022
		(0.089)		(0.096)
SIDS		0.624***		0.754***
		(0.095)		(0.115)
ND-GAIN adapt. capacity		−0.315		−1.141***
		(0.294)		(0.323)
GDP per capita		1.752***		1.819***
		(0.432)		(0.457)
GDP per capita sq.		−0.138***		−0.156***
		(0.028)		(0.029)
WGI index		0.830***		0.822***
		(0.066)		(0.072)
Exports			0.098***	0.127***
			(0.013)	(0.021)
Geographic distance			0.160***	0.028
			(0.036)	(0.043)
Colonial dummy			0.724***	0.818***
			(0.114)	(0.126)
UN joint voting			−0.057	−0.346**
			(0.127)	(0.164)
Total dev. aid	0.671***	0.686***	0.670***	0.616***
	(0.019)	(0.020)	(0.019)	(0.022)
Population	0.037**	0.200***	−0.017	0.095***
	(0.017)	(0.019)	(0.019)	(0.029)
Constant	−6.702***	−11.986***	−6.763***	−11.987***
	(0.489)	(1.783)	(0.458)	(1.925)
Observations	20,220	22,102	22,768	19,970
Log likelihood	−4938.680	−5159.981	−5394.652	−4692.484
Akaike inf. crit.	9901.360	10,351.960	10,815.310	9430.969
Bayesian inf. crit.	9996.333	10,480.020	10,919.740	9612.714

Note: year fixed and donor random effects omitted
Standard errors in parentheses
$^{*}p < 0.1$; $^{**}p < 0.05$; $^{***}p < 0.01$

Table 5.2 Principal adaptation aid, allocation stage, all countries

	Dependent variable: amount of principal adaptation aid			
	(1)	(2)	(3)	(4)
ND-GAIN exposure	0.621***			0.532***
	(0.172)			(0.183)
EVI	0.087***			0.037**
	(0.014)			(0.016)
CRI	−0.0001			0.00004
	(0.0003)			(0.0003)
African dummy		0.075***		0.065**
		(0.028)		(0.026)
LDCs		0.114***		0.073**
		(0.035)		(0.032)
SIDS		0.222***		0.193***
		(0.041)		(0.041)
ND-GAIN adapt. capacity		−0.575***		−0.381***
		(0.122)		(0.118)
GDP per capita		0.493***		0.345**
		(0.177)		(0.159)
GDP per capita sq.		−0.029***		−0.022**
		(0.011)		(0.010)
WGI index		−0.002		0.015
		(0.028)		(0.025)
Exports			0.045***	0.021***
			(0.006)	(0.007)
Geographic distance			0.020	0.012
			(0.013)	(0.012)
Colonial dummy			−0.196***	−0.162***
			(0.041)	(0.036)
UN joint voting			−0.023	0.053
			(0.055)	(0.053)
Total dev. aid	0.080***	0.108***	0.096***	0.092***
	(0.007)	(0.008)	(0.008)	(0.007)
Population	−0.130***	−0.143***	−0.208***	−0.130***
	(0.006)	(0.007)	(0.008)	(0.010)
Constant	1.551***	0.500	2.468***	0.051
	(0.149)	(0.730)	(0.159)	(0.675)
Observations	2357	2474	2509	2343
Log likelihood	−1431.960	−1812.408	−1847.524	−1387.767
Akaike inf. crit.	2889.920	3658.817	3723.047	2823.534
Bayesian inf. crit.	2964.867	3757.648	3804.634	2961.754

Note: year fixed and donor random effects omitted
Standard errors in parentheses
$^{*}p < 0.1$; $^{**}p < 0.05$; $^{***}p < 0.01$

Table 5.3 Principal and discounted significant adaptation aid, selection stage, all countries

	Dep. variable: total adaptation aid—Yes/No			
	(1)	(2)	(3)	(4)
ND-GAIN exposure	2.649***			3.323***
	(0.391)			(0.472)
EVI	−0.190***			−0.039
	(0.032)			(0.040)
CRI	0.008***			0.005***
	(0.001)			(0.001)
African dummy		−0.309***		−0.468***
		(0.058)		(0.065)
LDCs		−0.046		−0.174**
		(0.076)		(0.082)
SIDS		0.213***		0.238**
		(0.078)		(0.095)
ND-GAIN adapt. capacity		0.206		−0.473*
		(0.242)		(0.266)
GDP per capita		0.757**		0.894**
		(0.339)		(0.360)
GDP per capita sq.		−0.073***		−0.094***
		(0.022)		(0.023)
WGI index		0.788***		0.844***
		(0.053)		(0.058)
Exports			0.065***	0.108***
			(0.010)	(0.016)
Geographic distance			0.129***	−0.085**
			(0.031)	(0.039)
Colonial dummy			0.699***	0.781***
			(0.123)	(0.136)
UN joint voting			−0.005	−0.507***
			(0.107)	(0.140)
Total dev. aid	0.956***	0.949***	0.936***	0.896***
	(0.020)	(0.020)	(0.019)	(0.021)
Population	0.095***	0.228***	0.081***	0.109***
	(0.014)	(0.016)	(0.015)	(0.024)
Constant	−6.075***	−8.087***	−6.755***	−6.939***
	(0.467)	(1.405)	(0.441)	(1.517)
Observations	20,220	22,102	22,768	19,970
Log likelihood	−6612.165	−6993.042	−7374.402	−6323.513
Akaike inf. crit.	13,248.330	14,018.080	14,774.800	12,693.030
Bayesian inf. crit.	13,343.300	14,146.140	14,879.230	12,874.770

Note: year fixed and donor random effects omitted
Standard errors in parentheses
$^{*}p < 0.1$; $^{**}p < 0.05$; $^{***}p < 0.01$

Table 5.4 Principal and discounted significant adaptation aid, allocation stage, all countries

	Dep. variable: amount of total adaptation aid			
	(1)	(2)	(3)	(4)
ND-GAIN exposure	0.510***			0.421***
	(0.111)			(0.119)
EVI	0.068***			0.007
	(0.009)			(0.010)
CRI	0.0002			−0.00000
	(0.0002)			(0.0002)
African dummy		0.010		0.003
		(0.017)		(0.016)
LDCs		0.106***		0.086***
		(0.022)		(0.020)
SIDS		0.250***		0.229***
		(0.025)		(0.026)
ND-GAIN adapt. capacity		−0.304***		−0.257***
		(0.074)		(0.072)
GDP per capita		0.457***		0.377***
		(0.103)		(0.096)
GDP per capita sq.		−0.028***		−0.024***
		(0.007)		(0.006)
WGI index		0.052***		0.049***
		(0.016)		(0.015)
Exports			0.036***	0.023***
			(0.003)	(0.005)
Geographic distance			0.010	0.006
			(0.009)	(0.008)
Colonial dummy			−0.199***	−0.168***
			(0.030)	(0.028)
UN joint voting			0.019	0.037
			(0.036)	(0.036)
Total dev. aid	0.118***	0.140***	0.135***	0.122***
	(0.004)	(0.005)	(0.005)	(0.005)
Population	−0.145***	−0.139***	−0.203***	−0.135***
	(0.004)	(0.005)	(0.005)	(0.006)
Constant	1.798***	0.439	2.514***	0.123
	(0.102)	(0.421)	(0.109)	(0.401)
Observations	4986	5222	5318	4959
Log likelihood	−2718.712	−3248.775	−3376.647	−2539.552
Akaike inf. crit.	5463.425	6531.549	6781.294	5127.104
Bayesian inf. crit.	5548.112	6643.080	6873.398	5283.319

Note: year fixed and donor random effects omitted
Standard errors in parentheses
$^{*}p < 0.1$; $^{**}p < 0.05$; $^{***}p < 0.01$

5.1.1 *Physical Exposure and Sensitivity to Climate Change Impacts*

We first discuss how our three measures of physical exposure and sensitivity to climate change affect adaptation aid allocation. All variables related to hypothesis H1a are coded such that higher values indicate higher physical vulnerability. As more vulnerable countries should receive more adaptation aid (see the discussion in Chap. 2), we expect to find positive relationships according to the recipient need model.

Model 1 in Tables 5.1, 5.2, 5.3, and 5.4 include the three measures of physical vulnerability—the ND-GAIN exposure sub-index, the inverted Climate Risk Index (CRI), and the Environmental Vulnerability Index (EVI)—in addition to the two control variables: total development aid and population size. These first models can therefore be seen as partial (physical vulnerability) models. The full model (model 4) in Tables 5.1, 5.2, 5.3, and 5.4 includes the three measures of physical vulnerability as well as all other variables. When inspecting the partial and the full models in the four tables, we note that the results are in general fairly stable, particularly when using principal adaptation aid as the dependent variable. For the second dependent variable that includes discounted significant aid, the EVI loses significance from the partial to the full model at both stages, but the results remain stable otherwise. This overall robustness is an indication of the stability and validity of the results reported in the tables.

Figure 5.2 graphically presents the effects of the three measures of physical vulnerability of the full model for our first dependent variable, principal adaptation aid per capita. The figure has six plots: the first three plots in the upper row (labelled a–c) show the results for the selection stage. Consequently, the *y*-axis of these plots represents the probability that countries receive adaptation aid when—all else being equal—the independent variables are set to a specified value. The plots in the lower row (labelled d–f) show the results for the same three independent variables for the allocation stage. The *y*-axis in these plots therefore denotes the amount of allocation aid countries are predicted to receive at specified values of the independent variables, conditional on receiving some adaptation aid at the selection stage and all else being equal. To construct the predicted values at both stages, we set all continuous confounders in the model to the mean, and the four dummy variables in the model (African countries, LDCs, SIDS, colony dummy) to zero. In the following subsections, we look at the effects of the three measures of physical exposure and sensitivity separately, and explain how they affect—according to our regression models—the

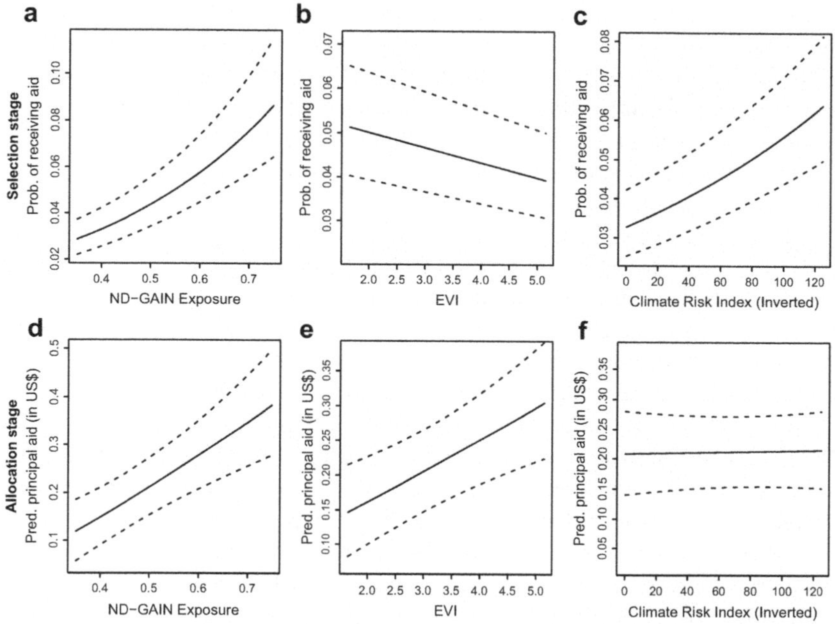

Fig. 5.2 Effects of the physical exposure variables for the full principal aid models of the selection stage (above) and the allocation stage (below), including 95% confidence intervals

probability that countries receive adaptation aid at the selection stage and the amount of adaptation aid they obtain at the allocation stage. We then summarise how these findings relate to H1a and how they compare to other studies of adaptation aid and the wider literature on aid allocation.

ND-GAIN Exposure

We start by interpreting the ND-GAIN exposure sub-index (ND-GAIN n.d.), which has by far the strongest effect of all physical vulnerability variables for both dependent variables and at both aid allocation stages. Remember that higher values denote higher physical exposure to climate change. Panel (a) of Fig. 5.2 shows that the probability of a country receiving some adaptation aid increases strongly as its physical exposure rises. For the countries with the lowest physical exposure values, the full model in Table 5.1 predicts a probability of less than 3% of receiving any adaptation aid at all. For the most exposed countries (as measured by the ND-GAIN sub-index), this probability of receiving adaptation aid increases more than threefold to almost 10%.[1] The effect of the ND-GAIN exposure sub-index is highly significant at the 99% confidence level in all our models. Overall, this is a strong effect and shows that more vulnerable countries—as measured by the ND-GAIN exposure sub-index—are more likely to be selected as recipients of adaptation aid.

Physical exposure as measured by the ND-GAIN exposure sub-index not only increases the likelihood of receiving some adaptation aid, but also increases the amount of adaptation aid per capita that selected countries can expect to receive.[2] Panel (d) of Fig. 5.2 again shows a clear upwards trend: the full model of Table 5.2 predicts a clear increase in the amount of principal adaptation aid as physical exposure increases. In the regression models, we used the natural logarithm of the dependent variable to account for the strong skewness of the measure. In panel (d) of the figure (and all other panels depicting allocation stage effects), we transformed the predicted values back to dollar terms to facilitate interpretation. When physical exposure as reported by ND-GAIN is very low, the selected countries are predicted to receive around 12 cents per capita for principal adaptation. This value increases to over 39 cents for the most exposed countries. In other words, the amount of adaptation aid per capita a developing country can expect to receive—conditional on receiving some adaptation aid—more than triples from the lowest to the highest physical exposure values.

At this point, let us take a closer look at the predicted per capita figures reported in the last paragraph. Recall that these predicted values only apply to countries that have been selected as adaptation aid recipients at the selection stage. From this point of view, the predicted values might—at first glance—seem very small, given that the average recipient country in our dataset can expect to receive approximately 37 cents per year for principal adaptation projects from *all* donors combined. However, the predictions of the models must be interpreted from a dyadic perspective, that is, if a recipient was selected by a *single* donor in a given year, then the model predicts adaptation aid flows per capita *from that specific donor*. As should be expected, the contributions of individual donors are quite limited, yet recipient countries can receive adaptation aid from more than one donor and in more than 1 year. Thus, the amount of principal adaptation aid per capita that a recipient country may receive in a given year can reach much higher levels than the ones depicted in panel (d) of Fig. 5.2. The same discussion of course also applies to all other allocation stage effects reported in the various models, and also for the models using principal and discounted significant adaptation aid.[3]

When turning to the models using principal and discounted significant adaptation aid as the dependent variable reported in Tables 5.3 and 5.4, we see that the coefficients of the ND-GAIN exposure sub-index are again highly significant for both stages in the partial and full models. The full model of the selection stage for this dependent variable predicts that the least exposed countries—all else being equal—have a probability of around 9% of being selected as adaptation aid recipients. For the most exposed countries, the probability of being selected again more than triples to almost 28%. This clearly shows that physical exposure increases quite dramatically the probability that countries receive some adaptation aid.

While the amount of principal and discounted significant adaptation aid is of course higher than the amount of principal adaptation aid only, it is also more widely spread across a larger number of recipients. Therefore, a recipient country—once selected—cannot expect much more support per capita for principal as well as significant adaptation projects than what it can expect to obtain for principal adaptation projects only. For the least exposed countries, the full model of Table 5.4 predicts on average 19 cents per capita for principal and significant adaptation projects[4] (about 7 cents more than for principal adaptation only). For the most exposed countries, the full model predicts on average 42 cents per capita (only about 2 cents more than for principal adaptation). In other words, significant adaptation aid tends to be distributed among a much larger set of recipient countries, yet the most vulnerable countries cannot necessarily expect to get a much larger piece of the pie. Among this larger set of countries receiving principal and significant adaptation aid, physical exposure as reported by ND-GAIN remains important, both for the selection and allocation stage. Yet when we include significant adaptation aid, the level of adaptation aid per capita remains rather stable for highly exposed countries, while it increases for less exposed countries. This indicates that projects where adaptation is only a significant objective do indeed also have goals different from climate change adaptation. To some extent, this can be seen as an indication of the reliability of the data reported to the OECD.

Overall, the findings for the ND-GAIN exposure sub-index combined are a first indication for the validity of our H1a—namely that more physically vulnerable countries receive more adaptation aid.

Environmental Vulnerability Index

Having discussed the ND-GAIN exposure variable in some detail, we now turn to the EVI, our second measure of physical vulnerability (see Kaly et al. 2004). As can be seen from Tables 5.1, 5.2, 5.3, and 5.4, the results for this variable are much weaker than for the ND-GAIN exposure sub-index. At the selection stage, the effects are mostly insignificant—except for the partial model of the selection stage using both principal and significant aid (model 1 in Table 5.3), where the result even indicates a negative effect between the EVI and adaptation aid. We therefore conclude that this measure of vulnerability does not have an impact on how donors decide as to which recipients to give adaptation aid. In contrast, the effect of the EVI is mostly significant at the allocation stage and has the expected sign. In other words, more vulnerable countries—as measured by the EVI—tend to receive more adaptation aid per capita, conditional on receiving some adaptation aid. Yet, the effect is insignificant in the full model for principal and discounted significant adaptation aid (model 4 in Table 5.4). Overall then, the results for the EVI suggest that more vulnerable countries get more adaptation aid, but the evidence is relatively weak.

Panels (b) and (e) of Fig. 5.2 display the effect of the EVI for principal adaptation aid for the full models at the selection and allocation stages. As can be seen, we find a slightly negative but insignificant effect of the indicator at the selection stage, as already pointed out. We therefore focus the discussion here on the allocation stage, where higher EVI vulnerability values are associated with higher predicted amounts of adaptation aid per capita. As the figure shows, from the lowest to the highest EVI values, the predicted principal adaptation aid per capita increases from just below 15 cents to just over 30 cents per capita (all else being equal and conditional on receiving some adaptation aid at the selection stage). While this effect is weaker than the one observed for the ND-GAIN exposure sub-index, the amount of adaptation aid recipients can expect to obtain still more than doubles from the least to the most vulnerable countries.

One reason for the comparatively weak effects of the EVI compared to ND-GAIN exposure sub-index might relate to the age of the EVI. The EVI was first published in 2004 and has not been updated since. While some physical characteristics such as land area or relief hardly change over time, other components of the EVI such as dry and wet periods or human population density do change, such that an indicator of 2004 may be somewhat outdated and no longer play such an important role for adaptation aid allocation. Nonetheless, the EVI overall provides additional evidence in favour of our hypothesis H1a that more vulnerable countries receive more adaptation aid.

Climate Risk Index
Our final measure for physical vulnerability to climate change is the CRI published by Germanwatch (e.g. Kreft et al. 2016). As opposed to the ND-GAIN exposure sub-index and the EVI, this measure captures short-term impacts of weather-related extreme events, such as storms, floods, or heat waves. These events are highly visible and might influence allocation decisions directly, which is why we include the measure in our models. Another advantage of this measure is its annual availability: Germanwatch publishes annual CRI data. On the other hand, the weather events that underlie the CRI scores are not necessarily always related (or attributable) to climate change, as the providers of the index themselves emphasise (e.g. Kreft et al. 2016; see Chap. 3 for a more detailed discussion).

The CRI provides mixed evidence for our hypothesis that vulnerable countries are prioritised in adaptation aid allocation (H1a). While we find strong evidence in favour of H1a at the selection stage for both dependent variables, the coefficient for the CRI is insignificant at the allocation stage across all models reported in the tables. Exposure to weather related extreme events as measured by the CRI influences whether a country receives any adaptation at all, but does not seem to influence how much adaptation aid countries receive. We therefore focus the discussion here on the selection stage. Recall that we inverted the CRI, so that higher values indicate higher vulnerability.

The full models in Tables 5.1 and 5.3 show that higher CRI values are related to a higher likelihood of receiving adaptation aid (both principal only and principal and discounted significant adaptation aid). Panel (c) of Fig. 5.2 shows how the probability of receiving principal adaptation aid increases as vulnerability to climate risks—as measured by the CRI—increases. The increase, however, is not as strong as was the case with the ND-GAIN exposure sub-index. More specifically, the least vulnerable countries with the lowest inverted CRI scores have a predicted probability of around 3% of receiving adaptation aid. This value increases to about 6.5% for the most vulnerable countries—that is, the probability of receiving adaptation aid more than doubles from the least to the most vulnerable countries. Yet this effect is weaker than that of the ND-GAIN exposure sub-index, which predicted the probability of receiving principal adaptation aid to more than triple from under 3% to about 10%.

When we examine our second dependent variable that includes discounted significant adaptation aid (Table 5.3), the results also indicate that the ND-GAIN exposure sub-index is a stronger predictor of who receives adaptation aid. Again, the effect of the CRI is positive and highly significant.[5] The full model in Table 5.3 predicts that the probability of receiving adaptation aid is about 11% for the least vulnerable countries (as measured by the CRI). This value increases to about 19% for the most vulnerable countries. The effect of the CRI is weaker than that of the ND-GAIN exposure sub-index, which predicted the probability of receiving principal and discounted significant adaptation aid to increase from 9% to 28% from the least to the most exposed countries.

Overall, then, the models suggest that exposure to extreme weather events as measured by the CRI is an important driver of who receives adaptation aid, though not of how much adaptation aid a country receives. The findings for the CRI hence further confirm our expectation that more vulnerable countries receive more adaptation aid.

In summary, we find that one of our measures of physical vulnerability—the ND-GAIN exposure sub-index—is a very strong predictor of adaptation aid flows at both stages of adaptation aid allocation and for both dependent variables. The CRI provides additional evidence for physical vulnerability playing a role at the selection stage, as does the EVI at the allocation stage. In contrast to previous studies, our analysis hence suggests that recipient need—the 'heart' argument of David Cameron (see Lightfoot et al. 2016)—is a clear motive for adaptation aid allocation. Most studies on adaptation aid allocation only find a weak relationship, if any, between physical vulnerability and adaptation aid (Betzold 2015; Nakhooda et al. 2013; Robertsen et al. 2015; Robinson and Dornan 2017; Stadelmann et al. 2014), although some studies, including our own research (Barrett 2014; Betzold and Weiler 2017; Weiler et al. 2017), find evidence that physical vulnerability matters, in line with the present analysis.

As we mentioned in Chap. 2, there are many more indicators of (physical) vulnerability. It would be worthwhile to further test the relationship between physical vulnerability to climate change and adaptation aid using these other indicators. This would also allow us to understand more fully which aspects of vulnerability donors take into account. Yet given the results at hand, we conclude that there is reasonably convincing evidence in favour of the argument that donors do consider recipients' vulnerability—

at least its physical component—when distributing adaptation aid at both aid allocation stages. We therefore consider H1a to be substantiated: at least according to our analysis, donors seem to keep their promise of prioritising physically vulnerable countries.

5.1.2 *Adaptive Capacity and Good Governance*

How about the second component of vulnerability to climate change, adaptive capacity (H1b)? In this section, we discuss how adaptive capacity (H1b) as well as good governance (H2) affect the probability of getting adaptation aid at the selection stage and the amount of adaptation aid recipient countries receive at the allocation stage.

We discuss the two hypotheses regarding adaptive capacity and good governance together because disentangling these two determinants of adaptation aid allocation is difficult (see Chaps. 2 and 3). Higher levels of GDP per capita, for instance, indicate on the one hand higher adaptive capacity, since richer countries can generally better cope with the impacts of a changing climate. On the other hand, higher levels of GDP per capita also correlate with higher government efficiency, less wasteful resource use, and less corruption—indicators that are highly relevant in the context of good governance. In our dataset, GDP per capita correlates highly with the two variables of government efficiency (correlation coefficient of 0.57) and control of corruption (correlation coefficient of 0.45), as reported by the WGIs. Thus, GDP per capita can be understood as a measure of both adaptive capacity and good governance. The same logic also applies to the WGI composite index and the ND-GAIN adaptive capacity sub-index under investigation in this section. The six indicators of the WGI composite index (see Chap. 3 for details) all measure good governance, but better-governed countries also have a higher adaptive capacity. The ND-GAIN adaptive capacity sub-index—despite the name—can also be seen as a measure of good governance, as the variables it is composed of (such as access to sanitation and water, malnutrition of children, maternal mortality, or quality of infrastructure) not only capture how well a country is able to deal with climate change, but also how efficiently the government provides basic services and uses available resources, which are aspects of good governance.

Given these problems of separability, we discuss both our hypotheses H1b and H2 in this section. We have four measures of adaptive capacity and good governance. First, we consider GDP per capita, which is probably the most clear-cut measure of adaptive capacity (H1b). We then examine the vulnerable country dummies (African countries, LDCs, and SIDS) before turning our attention to the ND-GAIN adaptive capacity sub-index. Finally, we discuss the composite WGI index and particularly use the results of this measure to draw conclusions on our recipient merit (good governance) hypothesis (H2).

In general, we can formulate two expectations with regard to these measures of adaptive capacity and good governance: if donors pay more attention to countries' vulnerability and hence to their adaptive capacity, we should see a *negative* relationship: countries that have less adaptive capacity should receive more adaptation aid. In contrast, if donors are mainly concerned with countries' governance, we should see a *positive* relationship: countries that are better governed should receive more adaptation aid—even if they have a more adaptive capacity.[6]

Figure 5.3 graphically depicts the observed effects of our measures of adaptive capacity/good governance (except for the vulnerability dummies) for both stages of the full principal adaptation aid model (model 4 in Tables 5.1 and 5.2). As can be seen, our findings with regard to adaptive capacity and good governance are mixed. In general, they point to a fairly strong role of good governance and a more limited consideration of adaptive capacity. This finding is in line with previous research (e.g. Barrett 2014; Betzold and Weiler 2017; Weiler et al. 2017) and relates to the question of whether countries with low adaptive capacity are partly responsible for their vulnerability (see Füssel 2009). Nonetheless, we find some evidence that donors also consider adaptive capacity and allocate more adaptation aid to poor countries. We now discuss the results for the four measures of adaptive capacity/good governance in turn.

GDP Per Capita

We start our discussion with GDP per capita, or countries' financial capacity. Everything else being equal, we expect that the higher a country's GDP per capita, the higher its adaptive capacity and hence the less vulnerable the country. We should thus see a negative relationship between GDP per capita and adaptation aid. In practice, research indicates that the relationship between GDP per capita and aid is non-linear because of recipients' absorptive capacity: very poor countries receive less aid than we would expect because they are (perceived as) less able to use resources efficiently (see e.g. Alesina and Dollar 2000; Neumayer 2003). Accordingly, we also

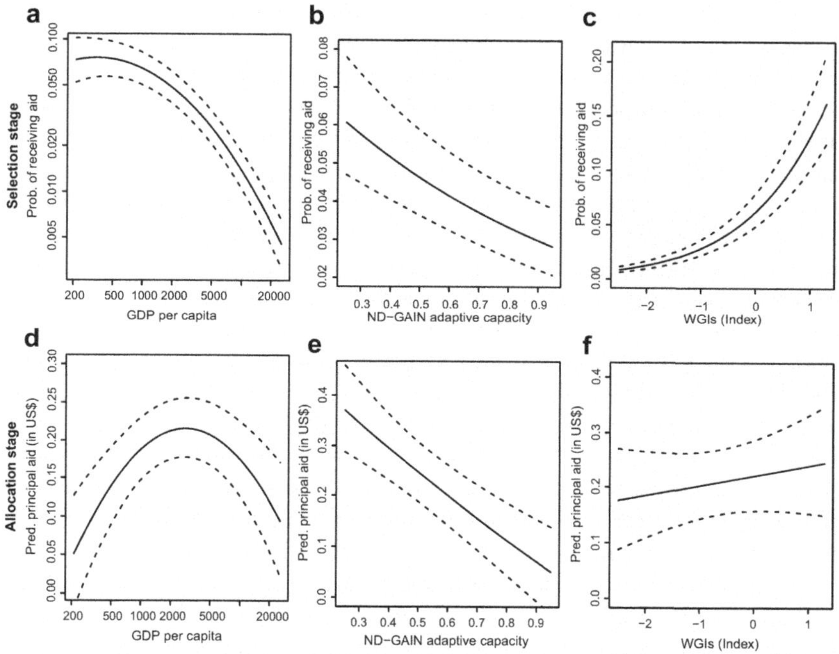

Fig. 5.3 Effects of the adaptive capacity and/or recipient merit variables for the full principal aid models of the selection stage (above) and the allocation stage (below), including 95% confidence intervals

include a squared term of GDP per capita in the regression models. The two coefficients capturing GDP per capita levels must be interpreted jointly to obtain the overall effect that GDP per capita has on adaptation aid allocation.

When we take a look at the relevant coefficients in Tables 5.1, 5.2, 5.3, and 5.4, we see that in all models the coefficient of GDP per capita is positive, while that of the squared term is negative. The effects are highly significant across the models at the 95% or 99% confidence level. In other words, when countries are very poor, both the probability of receiving adaptation aid and the amount of adaptation aid allocated increase as per capita income increases (as indicated by the positive coefficient of (not-squared) GDP per capita). However, as indicated by the negative squared term, this increase in the probability of receiving adaptation aid becomes smaller and smaller as GDP per capita increases, until the predicted probability reaches a peak and declines thereafter. The same applies for the level of adaptation aid per capita. In other words, very poor countries and very rich countries have a lower probability of receiving adaptation aid and at the same time also receive less adaptation per capita than middle-income countries (conditional on receiving some adaptation aid). We can interpret the increase of adaptation aid at very low levels of per capita income as donors prioritising good governance over recipient need: very poor countries, although vulnerable to climate change because of their low adaptive capacity, also tend to be poorly governed and donors therefore reluctant to give adaptation aid. On the other hand, we can interpret the decrease in the likelihood of receiving adaptation aid and the level of adaptation aid per capita at higher levels of GDP per capita as donors taking into account adaptive capacity: very rich countries, although well governed, have high adaptive capacity and therefore need less aid for adaptation. From this perspective, the findings for GDP per capita support both the recipient need and recipient merit arguments (hypotheses H1b and H2).

In order to evaluate more closely the two hypotheses in question, let us take a closer look at panels (a) and (d) of Fig. 5.3. As before, both figures are for the full models using principal adaptation aid only as the dependent variable. Panel (a) shows that the predicted probability of receiving adaptation aid (selection stage) only increases very slightly from the lowest GDP per capita values recorded in the dataset (at \$216) and reaches its peak at the only somewhat higher GDP per capita value of around \$350. The predicted probabilities of receiving aid for these two very low values of income are very similar, at 7.4% and 7.5%, respectively. The difference between these predicted probabilities is not statistically significant. When income increases further, however, the predicted probabilities of receiving adaptation aid decline relatively quickly. The probability of receiving adaptation aid drops to 6.4% at a GDP per capita of \$1000 and then further drops to just 0.6% at \$20,000. From the poorest to the richest countries, the predicted probability thus decreases more than tenfold. We see a very similar pattern when we look at the selection stage of the full model in Table 5.3 with principal and discounted significant adaptation aid as the dependent variable (not graphically depicted). Here, the predicted probability of receiving adaptation aid starts at 27.6% at the lowest recorded level of GDP per capita. This is also the maximum predicted value since the peak is outside the observed value range. It then drops to

24.4% at $500 and to 3.2% at $20,000. Again, this represents a dramatic decline in the predicted probability of receiving adaptation aid from the poorest to the richest countries. Overall, the downward trends clearly show that adaptive capacity (and thus recipient need) plays an important role when donors make decisions on who receives adaptation aid (H1a). The capacity of governments to use resources efficiently (H2), on the other hand, does not seem to play such a large role at the selection stage—very poor countries are not significantly less likely to receive adaptation aid than poor countries.

How about the allocation stage? Here, the story is somewhat different, as panel (d) of Fig. 5.3 clearly shows.[7] At this second stage, the poorest countries—once selected to receive some adaptation aid—receive significantly *less* adaptation aid than middle-income countries (and particularly lower-middle-income countries). As the figure shows, inhabitants of a very poor country with a per capita GDP of around $250 per capita (such as Malawi, Burundi, or the Central African Republic), are predicted to obtain only about 7 cents for principal adaptation projects from an individual donor who decided to support adaptation in this country. When GDP per capita doubles to $500, the predicted amount of principal adaptation aid per capita also doubles to 14 cents. Another doubling of GDP per capita to $1000 increases the predicted value of adaptation aid to 19 cents, which is already quite close to the peak of 21 cents reached at a GDP per capita of around $2500. Thereafter, the amount of adaptation aid per capita again gradually declines as GDP per capita increases. A country with a GDP per capita of about $10,000 can expect to receive around 17 cents per inhabitant, while the richest countries in the dataset who receive some adaptation aid are predicted to collect about 9 cents per capita for principal adaptation. Thus, the richest countries are predicted to receive more adaptation aid (on a per capita basis) than the poorest countries in the dataset. Yet it should be stressed again that these figures are contingent on countries receiving some adaptation aid and represent adaptation aid from individual donors (and that recipients receive adaptation aid from more than one donor). The richest countries have a much lower probability of being recipients of adaptation aid than the poorest countries in our dataset. Moreover, there are only a few countries with GDP per capita values below $500, at which point—according to the models—countries already receive more principal adaptation aid per capita than countries with GDP per capita values above $15,000. The picture is similar when we use principal and discounted significant adaptation aid, if less favourable for rich economies. While inhabitants of the poorest countries here are predicted to receive about 13 cents for adaptation, this value already increases to 20 cents for countries with a GDP per capita of $500. The peak in this model is at around $2000, with predicted per capita adaptation aid levels of around 26 cents. Predicted adaptation aid per capita then drops to 19 cents at GDP per capita of $10,000, and further drops to just 9 cents for the richest countries in the dataset (again, all conditional on receiving some adaptation aid).

Considering the two stages of adaptation aid allocation combined, then, we conclude that there is strong evidence for a preferential treatment of poorer and more vulnerable countries. Poorer countries are more likely to receive adaptation aid (selection stage) and also receive more adaptation aid per capita (allocation stage) than richer countries. This should not come as a surprise, since adaptation aid is part of ODA which by definition aims to promote economic development and welfare (see Chap. 1). Our qualitative analysis similarly suggests that donors recognise the links between poverty and vulnerability and acknowledge that poor countries need particular support with adaptation to climate change (see Chap. 6). Nonetheless, we also find that very poor countries are (slightly) less likely to receive adaptation aid and receive less of it per capita. This finding suggests that, beyond recipient need, aid effectiveness matters: donors do not only consider recipient needs when allocating adaptation aid, but also seek to maximise the impact of their aid. In very poor countries with little absorptive capacity, aid is probably less effective than in lower-middle-income countries with more developed institutions. Overall, however, the findings for GDP per capita provide support for the recipient need argument (H1b) rather than for the recipient merit argument (H2), and thus confirm our earlier conclusion that donors do take into account vulnerability when allocating adaptation aid.

Particularly Vulnerable Countries

The next measures we discuss are our three vulnerable country dummies for African countries, LDCs, and SIDS. As discussed in Chap. 3, these dummy variables are not a pure measure of countries' adaptive capacity. Rather, these groups of countries have been singled out as 'particularly vulnerable' in the climate change negotiations because they are highly exposed and sensitive to climate change and at the same time have little capacity to cope with and adapt to climate change impacts. In particular the SIDS dummy also incorporates a measure of physical vulnerability, as SIDS are particularly exposed and sensitive to environmental changes, including climate change (see Nurse et al. 2014). LDCs, in contrast, tend to have a very low adaptive capacity, while African countries—where many LDCs are located—fall somewhere in between: 'Africa as a whole is one of the most vulnerable continents due to its high exposure *and* low adaptive capacity' (Niang et al. 2014, 1205; emphasis added) Yet, these are broad generalisations; we recognise that the three groups of particularly vulnerable countries are very heterogeneous and partly overlap.

Starting with the selection stage, we can see in Tables 5.1 and 5.3 that SIDS are predicted to have a higher probability of being selected as adaptation aid recipients than the average country. For African countries, on the other hand, the opposite is the case: the models predict that these states have a *lower* probability of receiving adaptation aid than non-African countries. The effect for LDCs, while insignificant for principal aid only (Table 5.1), is also negative and significant in the full model of Table 5.3 for principal and discounted significant adaptation aid. The data thus suggest that these very poor and vulnerable countries have a lower probability of receiving adaptation aid.

Figure 5.4 plots the effects of the three dummy variables for the full model for our first dependent variable that only considers principal adaptation aid (model 4 in Table 5.1). Panel (a) displays the predicted probability of receiving adaptation aid for the three country groups compared to other countries. Panel (b) shows the predicted amount of principal adaptation aid per capita for the three country groups compared to other countries.

As panel (a) shows, the baseline probability of receiving principal adaptation aid when all vulnerable country dummies are set to zero is 6.5%. The probability of LDCs receiving adaptation aid is 6.6%, only insignificantly higher than the baseline probability. In contrast, the probability of African countries receiving adaptation aid is lower than the baseline at 4.1%, while that of SIDS is much higher at 12.9%. When we consider discounted significant adaptation aid as well, the baseline probability of a country receiving adaptation aid is much higher than for principal adaptation aid alone: 17.1% (compared to 6.5%). Yet, the probability for African countries and for LDCs receiving adaptation aid is below the baseline probability, at 11.4% and 14.7%, respectively. In contrast, SIDS are expected to receive adaptation aid in 20.7% of cases. Overall, these results again suggest that physical vulnerability—as in the SIDS case—plays an important role in how donors decide as to which countries they want to support adaptation at the selection stage (and thus again support H1a). The negative and

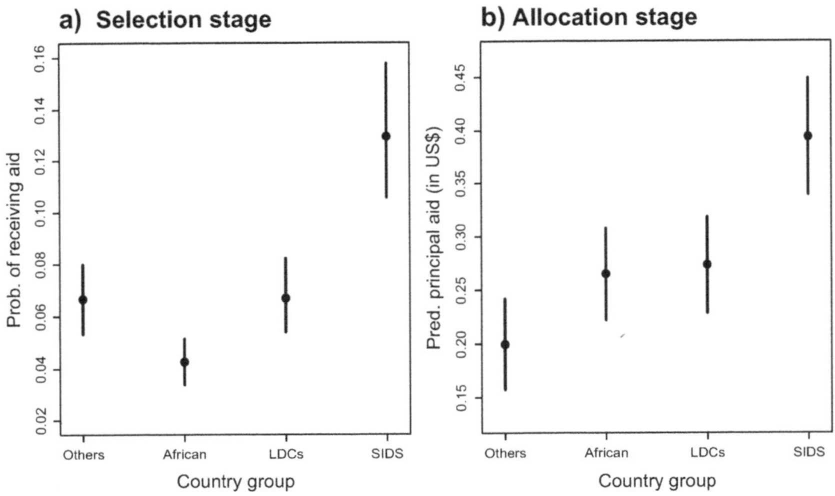

Fig. 5.4 Effects of the vulnerability country dummies for the full principal aid models of the selection stage (panel (**a**)) and the allocation stage (panel (**b**)), including 95% confidence intervals

partly significant effects for African countries and LDCs is more puzzling. We would interpret this finding not as an indication that donors do not take into account low adaptive capacity when allocating adaptation aid. Rather, the results may suggest that African countries and LDCs have many other, seemingly more pressing, challenges. Development projects in these countries may have first to address issues related to capacity, infrastructure, or health before they can turn to (longer-term) adaptation—which is also what our interviewees suggested (see Chap. 6).

The results for the allocation stage are much stronger and indicate that particularly vulnerable countries do get more adaptation aid—once they have been selected to receive some of it. For principal adaptation aid only, the full model of Table 5.2, shown in panel (b) of Fig. 5.4, predicts a baseline of 19 cents per capita for countries that do not belong to any of the three groups. Inhabitants of African countries, in contrast, can expect—on average—to receive 26 cents, those of LDCs 27 cents, and those of SIDS 39 cents for principal adaptation projects—all conditional on receiving some adaptation aid at the selection stage. For principal and discounted significant adaptation aid, reported in Table 5.4, the respective numbers are 24 cents per capita for African countries, 32 cents for LDCs, and 46 cents for SIDS, compared to 23 cents for baseline countries. Note that the effect for African countries is not significant in this model.

Overall, then, it seems that particularly vulnerable country status does affect how much adaptation aid countries can expect to receive—once donors have decided to provide adaptation aid. The results for both the selection and in particular the allocation stage support our expectation that vulnerability matters; vulnerable countries—be it because of their high exposure and sensitivity (H1a), be it because of their low adaptive capacity (H1b)—are prioritised in adaptation aid allocation.

ND-GAIN Adaptive Capacity

We now take a closer look at our third measure of adaptive capacity/good governance: the ND-GAIN adaptive capacity sub-index. This sub-index is quite ambiguous in terms of whether it captures adaptive capacity or good governance. As mentioned earlier, this sub-index consists of a range of variables such as access to sanitation and water, malnutrition of children, maternal mortality, and quality of infrastructure, which on the one hand capture dimensions of adaptive capacity, but on the other hand also depend on governmental quality and competence. We therefore posit two contrasting expectations related to this variable. Remember that the ND-GAIN adaptive capacity sub-index is coded such that higher values indicate *lower* adaptive capacity (see Chap. 3). Thus, if the ND-GAIN adaptive capacity sub-index signals mainly vulnerability—as intended by the organisation providing the measure[8]—we would expect a *positive* relationship with adaptation aid: more vulnerable countries get more adaptation aid. If, in contrast, the ND-GAIN adaptive capacity sub-index signals recipient merit—that is, good governance—we would expect a *negative* relationship with adaptation aid: better governed and thus less vulnerable countries get more adaptation aid.

Our models find a negative and (mostly) significant effect across both stages of adaptation aid allocation and for both dependent variables under investigation. Thus, at *lower* ND-GAIN adaptive capacity scores, that is, when countries are better governed, they are more likely to be selected as adaptation aid recipients, and they receive higher amounts of this aid per capita at the allocation stage—although better governed countries are less vulnerable to climate change and thus less in need of external support.

This result, then, indicates that the ND-GAIN adaptive capacity sub-index mainly captures good governance rather than vulnerability.

When we take a closer look at the selection stage of adaptation aid allocation in the principal adaptation aid only model (see Table 5.1 and panel (b) of Fig. 5.3), we find that from the lowest to the highest ND-GAIN adaptive capacity scores (in other words, from lowest to highest vulnerability), the predicted probability of receiving adaptation aid more than halves, from 6.2% to 2.9%. Although this effect is not as strong as others reported in this chapter, it is nevertheless significant and indicates that donors do consider recipient merit when selecting which countries are to receive adaptation aid. The full model including principal and discounted significant adaptation aid (model 4 in Table 5.3) confirms this interpretation. It predicts that the probability of receiving adaptation aid falls from 16.3% for countries with the highest ND-GAIN adaptive capacity scores (best governance and lowest vulnerability) to about 12.5% for those with the highest ND-GAIN adaptive capacity scores (worst governance and highest vulnerability). This drop is weakly significant at the 90% confidence level.

The allocation stage indicates a similar pattern: developing countries with lower ND-GAIN adaptive capacity scores (lower vulnerability) get *less* adaptation aid, as panel (e) of Fig. 5.2 shows. However, the effect size is much larger at this second stage of adaptation aid allocation. Countries with the lowest ND-GAIN adaptive capacity scores—the least vulnerable and best governed countries—are predicted to receive around 37 cents per capita for adaptation (conditional on receiving some adaptation aid). In contrast, countries with the highest ND-GAIN adaptive capacity scores—the most vulnerable and worst governed countries—can only expect to get around 5 cents per capita and at the same time are less likely to receive any adaptation aid. The effect for the full model using both principal and discounted significant adaptation aid is similar, albeit somewhat smaller in magnitude. Here, our models predict a drop in the amount of per capita adaptation aid from 38 cents for the least vulnerable and best governed countries to 15 cents for the most vulnerable and worst governed countries.

While the results for GDP per capita have already hinted at the role of good governance, the results for the ND-GAIN adaptive capacity sub-index very strongly indicate that donors take recipients' governance seriously. Donors provide more adaptation aid to well governed countries, presumably because these are (seen as) best able to make good use of the aid. Overall, then, the results for the ND-GAIN adaptive capacity variable provide evidence in favour of our hypothesis H2 but against our hypothesis H1b.

Worldwide Governance Indicators

Finally, we turn to the composite WGI index. While GDP per capita is probably the clearest measure of adaptive capacity, the composite WGI index is probably the clearest measure of good governance—to the extent that good governance and adaptive capacity can be seen as separate concepts. The composite WGI index consists of six governance variables (see Chap. 3), which makes it an ideal candidate for measuring good governance and thus recipient merit. However, increases in good governance—that is, improvements in areas such as control of corruption, government efficiency, or the rule of law—can also be understood as improving a country's adaptive capacity, since countries doing better on these dimensions can also be expected to react faster and in a more coherent manner to the challenges posed by climate change. This should be kept in mind when interpreting the results of this measure. Following the development aid literature we expect better governed countries to be perceived as more deserving by donors, and thus to receive adaptation aid more frequently and in larger quantities (see e.g. Alesina and Dollar 2000; Dollar and Levin 2006; Zanger 2000).

Considering first the selection stage, we see that the coefficient of the composite WGI index is positive and highly significant across our models in Tables 5.1 and 5.3. In other words, better governed countries, as measured by the composite WGI index, have—all else being equal—a higher probability of receiving adaptation aid. Panel (c) of Fig. 5.3 displays the predicted probabilities of the full model considering principal adaptation aid only. The model predicts the probability of receiving such aid from a given donor to be below 1% for the worst governed countries (which also have the least adaptive capacity and are hence more vulnerable). This predicted probability increases dramatically to around 16% for the best governed countries in the dataset (which are also less vulnerable). This effect is also very strong when we consider the second dependent variable that includes both principal and discounted significant adaptation aid. In this case, the worst governed countries are predicted to receive some adaptation aid with a probability of under 3%, while the best governed countries are predicted to receive some with a probability of around 40%—this corresponds to an increase by a factor of 13.

Once selected, however, better governed countries do not automatically receive more adaptation aid per capita. At the allocation stage of the full model of Table 5.2 the coefficient of the WGI index is not significant. This can also be seen in panel (f) of Fig. 5.3, which shows that the model predicts per capita adaptation aid flows of around 20 cents no matter how well governed countries are. Does this mean, then, that once a donor has decided to provide adaptation aid to a country—based on the country's governance, among other things—it no longer considers governance at the allocation stage? The model using principal aid only would suggest so, yet when we take a closer look at the model that uses principal and discounted significant adaptation aid, we see a different picture. Here, the effect is positive and statistically significant at the 99% confidence level. For total adaptation aid, good governance not only increases the probability of receiving some adaptation aid, but also partly explains how much of it a country receives. While the worst governed countries are predicted to receive only about 14 cents per capita for principal and significant adaptation projects, this figure almost triples to around 38 cents for inhabitants of the best governed countries. Thus, despite the insignificant effect at the allocation stage of the principal aid model, we find some evidence that good governance plays a role at the second stage of adaptation aid allocation. This conclusion holds in particular when we also consider the findings of the ND-GAIN adaptive capacity sub-index.

Together with the findings of the ND-GAIN adaptive capacity sub-index discussed earlier, the findings for the composite WGI index very strongly indicate that donors do consider whether their resources will be spent efficiently when selecting their partner countries and when deciding how much adaptation aid to provide to selected countries. Donors support adaptation in well governed countries where their adaptation aid is presumably more effective—even if this means that they may not support adaptation in the most vulnerable countries with the least adaptive capacity.

Overall, then, these results are rather strong evidence in favour of our hypothesis H2: donors provide more adaptation aid to better governed countries, although they also consider poverty, and therefore partly adaptive capacity, as predicted by H1b. Our findings on adaptive capacity and recipient merit are therefore in line with those reported in the development aid literature, which also largely concludes that governance matters (e.g. Dollar and Levin 2006; Hoeffler and Outram 2011; Younas 2008). Our findings are also partly in line with those reported in the adaptation aid literature, which finds no relationship between adaptive capacity and adaption aid and also suggests a rather weak link between physical vulnerability and adaptation aid (e.g. Barrett 2014; Betzold 2015; Remling and Persson 2015; Robertsen et al. 2015; Robinson and Dornan 2017). While we conclude that vulnerability matters, our findings by and large support a climate-centred understanding of vulnerability: donors focus on the physical dimension of vulnerability rather than on underlying socio-economic drivers of vulnerability—not least because it is controversial as to what extent countries with low adaptive capacity are (partly) responsible for their high vulnerability (Füssel 2009, 18f; see also Chap. 1).

5.1.3 Donor Interests

Let us turn to our last hypothesis, which focuses on donor interests (H3). We have four measures of donor interests: exports from donors to developing countries; geographical distance between donor–recipient pairs; voting similarity in the UN General Assembly; and a dummy indicating former colonial ties between donors and recipients. Note that these variables are dyadic, that is, they depend on a specific donor–recipient pair. Accordingly, we expect that a recipient that is more 'relevant' to a given donor—a recipient that imports more from this donor, that is geographically closer to this donor, that votes similarly to this donor, or that is a former colony of this donor—is more likely to receive adaptation aid at the selection stage, and to receive higher amounts of this form of aid at the allocation stage from that specific donor.

Past studies on adaptation aid allocation and development aid allocation in general found donor interests to be an important, if not the major, explanation of how donors allocate their aid (e.g. Alesina and Dollar 2000; Barrett 2014; Berthélemy 2006; Betzold 2015; Hoeffler and Outram 2011). Our analysis only partly confirms this strong role of donor interests compared to recipient need and recipient merit. While donors strongly prefer to provide adaptation aid to countries to which they export a lot, we do not find that they prioritise countries that are geographically closer and obtain inconclusive findings with regard to colonial ties and voting similarity in the UN General Assembly (see the graphical overview of significant effects in Fig. 5.5). As opposed to development aid more broadly, adaptation aid seems to be more strongly guided by recipient need and recipient merit considerations as compared to donor interests.

Trade Ties Between Donors and Recipients

Our first measure of donor interests is trade volumes, measured by exports from donors to recipients. Recall that we use exports from donors to recipients rather than total trade between country pairs because we believe that exports better reflect donor interests (see Chap. 3). According to H3, we expect that higher exports lead to both a higher likelihood of receiving adaptation aid and higher amounts of this form of aid. To what extent do we see such an effect in the models? The relevant coefficients in Tables 5.1, 5.2, 5.3, and 5.4 show that all export effects are positive and highly significant at the 99% confidence level. More important trading partners receive adaptation aid more often and get more of it per capita, as our hypothesis H3 predicts. Recall that we control for population size in our models. Hence, the results of our trade variable are not simply a reflection of the size of the trading partner. In order to make more sense of the effects, let us have a closer look at what higher export values—according to our models—mean for the two stages of adaptation aid allocation.

Panel (a) of Fig. 5.5 illustrates the selection stage effect for the full model using principal adaptation aid only (model 4 in Table 5.1). Note that the x-axis of panels (a) and (d) of Fig. 5.5 is on an exponential scale. We immediately see that the effect is sizeable: developing countries with strong trade connections to donors have a much higher probability of receiving adaptation aid. The figure only shows countries with donor exports of at least \$100,000. Below that level, the probability of receiving adaptation aid is very close to zero. Yet as exports from donors to potential recipients increase, the probability of receiving adaptation aid rises quite fast. The probability of receiving adaptation aid rises to 3.5% at export volumes of around \$1 million, to 4.7% at \$10 million, and to over 6.0% at \$100 million. This might already seem like a quite large trade volume, but there are almost 6000 UN Comtrade entries relevant for our dataset that record even higher trade flows. As trade volumes increase further, the probability of receiving adaptation aid rises even faster, as the ever steeper curve in panel (a) of Fig. 5.5 at higher export values indicates. When trade volumes reach \$1 billion, the model already predicts a probability of receiving adaptation aid of over 8.0%, and at \$10 billion, of over 10.0%. For the almost 40 entries with trade flows over \$100 billion, the model predicts a probability of receiving adaptation aid of over 14.0%. When we instead consider the full model of the selection stage using principal and significant adaptation aid (model 4 in Table 5.3), the pattern is very similar, but

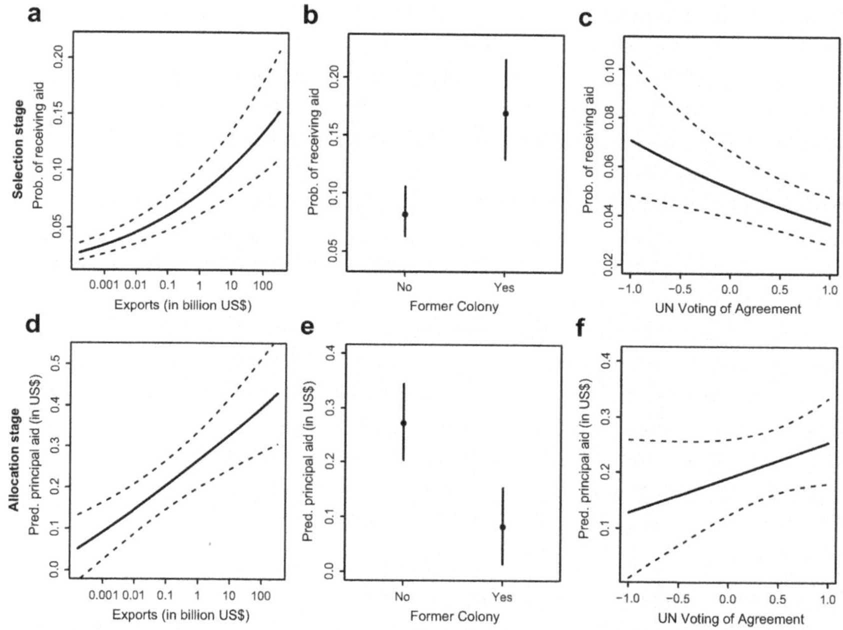

Fig. 5.5 Effects of the donor interest variables for the full principal adaptation aid models of the selection stage (above) and the allocation stage (below), including 95% confidence intervals

all predicted probabilities are much higher. The model thus predicts a probability of receiving adaptation aid of about 3% at no recorded trade flows, and then a more than tenfold increase to over 32% when export flows reach $100 billion.[9] Thus, for both dependent variables, the predicted probability of receiving adaptation aid increases dramatically as donor exports increase. Overall, this strongly indicates that donors' economic interests, as measured by exports from donor to recipient countries, play a substantial role in a donor's decision of whom to give adaptation aid.

The impact of trade relations on adaptation aid flows is also strong at the allocation stage, stronger than most other effects in the models. For the full model of our first dependent variable including principal adaptation aid only (model 4 in Table 5.2 and panel (d) of Fig. 5.5), we find a strong increase in the predicted amount of adaptation aid from the lowest to the highest levels of trade, from close to no adaptation aid at all to over 42 cents per capita for principal adaptation. When considering our second dependent variable including principal and discounted significant adaptation aid instead, the predicted amount of aid increases even more dramatically: countries to which donors do not export very much receive almost no adaptation aid from that donor, while the largest trading partners can expect to receive over 52 cents per capita (conditional on being selected at the first stage of aid allocation). There is no single other hypothesised effect in the models which impacts on aid flows as strongly at the allocation stage as this measure of donor interests.[10] As pointed out earlier, the same is true for the selection stage. Although we find that many other factors also contribute to whether recipients receive adaptation and, if so, how much, we must conclude that donors firmly consider their own economic interests when allocating adaptation aid. These results are thus in line with the findings of the development aid literature (see particularly Dudley and Montmarquette 1976; McKinlay 1978) as well as the adaptation aid literature (Barrett 2014; Betzold 2015), and strongly indicate the validity of our hypothesis H3.

Geographic Distance Between Donors and Recipients

Our second measure of donor interest is geographic distance between a donor and a recipient country, operationalised using R's 'cshapes' package. We expected countries that are closer to a donor to receive more adaptation aid, but we do not find a relationship between distance and adaptation aid. At the allocation stage, the coefficients are positive and insignificant across all models. At the selection stages, they are positive and significant in both partial models of Tables 5.1 and 5.3; positive but insignificant for the full model for principal adaptation aid only (model 4 in Table 5.1); and negative and significant for the full model for principal and discounted significant adaptation aid (model 4 in Table 5.3). Because these effects are too weak and too contradictory, they are not graphically depicted here. The positive effects in the partial models could simply indicate that most countries in need of adaptation aid are relatively distant geographically from donors, and the distance variable picks this up when not controlling for other factors. The insignificant effect at the selection stage of the full principal adaptation aid model and all models at the allocation stage indicate that geographical distance does not affect how donors allocate their adaptation aid. Against our expectations, donors do not prioritise countries that are geographically closer.

Colonial Ties Between Donors and Recipients

A third measure of donor interests is a dummy variable which indicates whether a donor was a former colonial power of a (potential) recipient country, using data provided by Hadenius and Teorell (2007), as reported in the Quality of Government dataset (Teorell et al. 2011). A priori, we would expect donors to keep political ties with their former colonies, including through providing adaptation aid. Yet, the results for the colonial dummy are unexpected. At the selection stage, the effects of the dummy variable are all positive and significant, while at the allocation stages they are negative—and again significant. In other words, former colonies are more likely to receive some adaptation aid from their former colonial powers. Yet, once selected, they receive significantly *less* per capita from their former colonial power than other countries. Apparently, donors deem it important to keep ties to former colonies and hence give them some adaptation aid, but other factors are more important regarding the level of adaptation aid that they provide. Since former colonies are over-represented in the population of recipient countries, they then receive lower levels of support at the allocation stage.

Taking a closer look at the selection stage for principal adaptation aid (panel (b) of Fig. 5.5), we observe that the probability of receiving adaptation aid is about twice as high for former colonies compared to other countries: the predicted probability for the former is 8.2% and for the latter 16.8%. Similarly, the full model of our second dependent variables including principal and discounted significant adaptation aid predicts a probability of receiving adaptation aid of 39.3% for former colonies but only of 22.8% for other countries. At the allocation stage, the findings indicate that a different logic of adaptation aid allocation is at play compared to the selection stage. Panel (e) of Fig. 5.5 shows the predicted level of aid for principal adaptation (full model). Countries that were not former colonies of a donor can on average expect to receive about 27 cents per capita from that donor. In contrast, former colonies only receive about 8 cents per capita for principal adaptation projects from their former colonial power (conditional on receiving some principal adaptation aid). These figures increase to 35 cents for non-colonies and 14 cents for former colonies when we also consider discounted significant adaptation aid projects.

Our conclusions regarding former colonial ties between donors and recipients are therefore ambivalent. Donors seem to keep political ties with their former colonies by allocating at least some adaptation aid to them, as the much higher probability at the selection stage shows. Yet actual adaptation aid flows from former colonial powers to their ex-colonies are lower than adaptation aid flows to other countries. What these findings mean for our hypothesis H3 is unclear.

Joint Voting in the United Nations General Assembly

Our fourth and final measure of donor interests is voting similarity in the UN General Assembly (Voeten et al. 2009). In line with hypothesis H3, we expect that donors provide more adaptation aid to countries that vote in a similar way, but we find no evidence that voting behaviour in the UN General Assembly is related to adaptation aid. At the allocation stage for both dependent variables under consideration, all coefficients of the voting variable are insignificant in both the partial and the full models. Only in the full models of the selection stage (again for both dependent variables) do we find significant effects. However, these effects are negative, which would indicate that recipients who vote more in line with a given donor are *less* likely to receive adaptation aid from that donor. This negative effect is significant at the 95% confidence level, but the effect size is not very large (see panel (c) of Fig. 5.5). The predicted probability of receiving principal adaptation aid from a given donor is 7% for countries that vote least often in line with that donor, but only 4% for countries that vote most often in line with the donor in question. Yet it should be noted that only about 7% of recipient countries in the dataset have similarity scores with any of the donors below zero (at which point they vote jointly in 50% of the cases), while almost 75% of country pairs have similarity values between 0.2 and 0.8, thus voting jointly in more than half the cases. Thus, it seems unlikely that donors pay much attention to voting behaviour in the UN General Assembly when allocating adaptation aid.

If we summarise our results for the four indicators of donor interests, we find that economic interests are very important: exports from a donor to a recipient is a very strong predictor of adaptation aid. In contrast, we find rather weak evidence for our other three measures of donor interests. While donors are more likely to give some adaptation aid to their former colonies, they provide less per capita to these countries. Finally, geographic distance and voting behaviour in the UN General Assembly do not seem to play any role in adaptation aid allocation. These results are rather surprising given the otherwise fairly strong evidence for donor interests in the aid allocation literature (e.g. Alesina and Dollar 2000; Berthélemy 2006; Dudley and Montmarquette 1976; McKinlay 1978). Nonetheless, we should not conclude that donor interests do not matter at all from our analysis, for at least two reasons. First, economic interests strongly matter: the effect of exports is the strongest hypothesised effect we find. Second, our measures of donor interests may be too broad to account for differences in how *adaptation* aid is allocated. For example, while voting in the UN General Assembly may not affect adaptation aid, negotiation behaviour in the UNFCCC climate change negotiations may indeed influence it, as our qualitative analysis hints at.

5.1.4 Control Variables

Before looking at the quantitative results for our three selected countries separately, we briefly turn to the control variables used in the statistical models: total development aid and recipient countries' population size. Both factors strongly influence adaptation aid allocation.

As can be seen in all statistical models (Tables 5.1, 5.2, 5.3, and 5.4), total development aid is a highly significant and very strong predictor of adaptation aid allocation, at both the selection and the allocation stage, and for both dependent variables. Note that the same also holds true in the models for our three selected countries (see next section and Tables 5.5 and 5.6), although the other effects are—in general—much weaker than in the pooled donor models. The full model of the selection stage in Table 5.1 predicts that countries receiving no to very little development aid from a specific donor also have an extremely low (very close to zero) probability of receiving adaptation aid from that donor. Countries that receive the highest amounts of development aid from a donor in contrast are very likely to receive adaptation aid from that donor, with a predicted probability of almost 80%. At the allocation stage, countries that receive only little development aid from a donor again can only expect to see very little adaptation aid coming their way (around 5 cents), while countries that receive the highest amounts are predicted to receive almost \$1 in adaptation aid from the respective donor. This clearly indicates the path dependency that adaptation aid has with respect to development aid: donors provide adaptation aid to countries with which they have an existing aid relationship (see Barrett 2014; Robertsen et al. 2015). This is not surprising as adaptation *aid* largely comes from development aid budgets. However, if donors use their adaptation *aid* to fulfil their adaptation finance commitments, the close link between adaptation aid and development aid becomes problematic. Donors have agreed to provide new and additional adaptation *finance* to vulnerable countries (see Chap. 2), and should respect this commitment when drawing on aid budgets to comply with international finance commitments. We return to this point in more detail in the section on implications, both from a policy and a research point of view, in Chap. 7.

Population, on the other hand, also has a strong and significant effect on adaptation aid at both stages of aid allocation, but this is much less problematic. Countries with large populations are much more likely to be selected as recipients of adaptation aid according to all models presented in this chapter (except for model 3 in Table 5.1). This indicates that donors are more likely to finance at least some adaptation-related projects in more populous countries. In simple terms, a given donor is more likely to give adaptation aid to India than to Bhutan. To some extent, this supports the donor interest argument, as larger countries also have more economic and political clout. In contrast, population size has a negative effect on the amount of per capita adaptation aid at the allocation stage—as the small country bias documented in the aid allocation literature predicts (see e.g. Alesina and Dollar 2000; Younas 2008). This should not be surprising, as the amount of adaptation aid that needs to flow to India would have to be over a thousand times that of Bhutan to reach the same per capita figure, as India's population is more than a thousand times that of Bhutan. Given that we only model the allocation stage for those countries that receive some adaptation aid at the selection stage, this negative effect is therefore expected and shows that it is important to control for population size.

5.2 Adaptation Aid Allocation in Germany, Sweden, and the UK

Now that we have a general idea of how donors—on average—distribute their adaptation aid, at both the selection and the allocation stage, we turn to Germany, Sweden, and the UK (see Chap. 3). As in the aggregate case, we run the models for the three countries in two stages. For the selection stage, which includes all potential recipient countries, we use logit models, since the dependent variable of this stage is binary: we want to understand which countries get adaptation aid regardless of the amount. For the allocation stage, we run standard ordinary least squares models, but only for those developing countries who did receive at least some adaptation aid from the donor in question at the selection stage. At both stages we again include year fixed effects. Thus, the only difference compared to the aggregate models is that we no longer include donor random effects, since the models only capture the allocation decisions of a single donor country.

Again, we use the, by now well known, two dependent variables—principal adaptation aid only and principal and discounted significant adaptation aid—to shed light on the adaptation aid allocation logics of the three selected countries. Table 5.5 shows the results for principal adaptation aid only for both the selection and the allocation stage, while Table 5.6 shows the respective findings for principal and discounted significant adaptation aid. In each table, columns 1–3 show the effects at the selection stage; columns 4–6 show the effects at the allocation stage.

A brief inspection of the models indicates that they are quite different from the aggregate models in that there are generally fewer significant effects, probably due to the now much smaller sample sizes. We further note that there are also many differences between the three selected donor models. This is not too surprising, since we selected donors that differ—at least to a degree—from each other based on a most different systems design. Such differences are also in line with the findings of the development aid literature, which show that individual donors vary widely in how they distribute their aid (e.g. Berthélemy 2006; Clist 2011). Thus, these results tell—to some extent—the story we expected: diverging decision-making processes and logics drive adaptation aid allocation in different donor countries. In addition, the results also show that individual donors may drive the results of single variables in the aggregate models, and that taking a closer look at individual donors is necessary to understand more fully adaptation aid distribution. We get back to this point in our qualitative analysis in the next chapter.

5.2.1 Germany

We start the discussion with Germany. As can be seen in Tables 5.5 and 5.6, there are some interesting similarities and differences between the models for the two dependent variables. Physical exposure and sensitivity to climate change does not seem to play a large role for Germany. The ND-GAIN exposure sub-index, which is a strong predictor of adaptation aid for all donors combined at both stages, does not affect German adaptation aid at all. The other two measures of physical vulnerability—the CRI and EVI—

Table 5.5 Selection and allocation stage for Germany, Sweden, and the UK, principal adaptation aid only

	Dependent variable: principal adaptation aid					
	Selection stage (*logit*)			Allocation stage (*OLS*)		
	GER	SWE	UK	GER	SWE	UK
ND-GAIN exposure	0.691	1.581	5.381**	−0.723	0.324	0.096
	(2.307)	(4.075)	(2.177)	(0.556)	(0.614)	(0.110)
EVI	−0.601***	−0.734**	−0.102	0.009	−0.152**	0.001
	(0.216)	(0.323)	(0.201)	(0.051)	(0.071)	(0.009)
CRI	0.010***	0.015**	0.002	0.0003	0.001	−0.0001
	(0.004)	(0.007)	(0.004)	(0.001)	(0.001)	(0.0002)
African dummy	−0.077	0.042	−0.821**	−0.002	0.059	−0.023
	(0.312)	(0.532)	(0.340)	(0.076)	(0.126)	(0.016)
LDCs	−0.090	−1.964***	0.118	0.127	−0.320***	−0.005
	(0.381)	(0.667)	(0.362)	(0.090)	(0.114)	(0.018)
SIDS	0.409		0.527	0.019		−0.020
	(0.530)		(0.443)	(0.134)		(0.023)
ND-GAIN adapt. capacity	−0.392	5.410**	−2.438**	0.132	−0.683	0.033
	(1.184)	(2.407)	(1.208)	(0.314)	(0.621)	(0.064)
GDP per capita	5.021***	4.226	1.678	1.004**	0.349	−0.158
	(1.663)	(3.503)	(1.791)	(0.406)	(0.647)	(0.106)
GDP per capita sq.	−0.383***	−0.295	−0.157	−0.057**	−0.031	0.010
	(0.110)	(0.228)	(0.114)	(0.026)	(0.044)	(0.007)

WGI index	0.698**	2.908***	1.091***	0.060	−0.088	−0.002
	(0.291)	(0.550)	(0.286)	(0.068)	(0.102)	(0.013)
Exports	0.270	−0.528***	0.188	−0.015	−0.006	0.002
	(0.170)	(0.152)	(0.117)	(0.048)	(0.040)	(0.007)
Geographic distance	0.988***	−0.875**	1.040***	−0.125**	−0.029	0.002
	(0.238)	(0.407)	(0.323)	(0.058)	(0.084)	(0.018)
Colonial dummy			0.199			0.001
			(0.302)			(0.014)
UN joint voting	1.733**	−1.583	−1.081	−0.627***	−0.126	0.029
	(0.817)	(1.321)	(0.779)	(0.201)	(0.269)	(0.045)
Total development aid	0.368***	0.935***	0.273***	0.138***	0.118***	0.016***
	(0.107)	(0.165)	(0.103)	(0.025)	(0.039)	(0.006)
Population	0.385**	0.950***	0.525***	−0.185***	−0.061	−0.017**
	(0.179)	(0.228)	(0.150)	(0.043)	(0.041)	(0.008)
Constant	−37.790***	−18.998	−25.231***	0.284	1.146	0.793*
	(7.420)	(13.869)	(8.064)	(1.800)	(2.359)	(0.428)
Observations	720	720	720	246	54	214
R^2				0.369	0.636	0.262
Adjusted R^2				0.313	0.433	0.182
Log likelihood	−300.639	−119.974	−295.278			
Akaike inf. crit.	643.279	279.948	634.556			

Note: year fixed effects omitted
Standard errors in parentheses
$^*p < 0.1$; $^{**}p < 0.05$; $^{***}p < 0.01$

Table 5.6 Selection and allocation stage for Germany, Sweden, and the UK, principal and (discounted) significant adaptation aid

	Dependent variable: principal and significant adaptation aid					
	Selection stage (*logit*)			Allocation stage (*OLS*)		
	GER	SWE	UK	GER	SWE	UK
ND-GAIN exposure	1.693	3.098	4.472**	−0.268	−0.010	0.401**
	(2.506)	(2.531)	(2.022)	(0.446)	(0.285)	(0.157)
EVI	−0.464**	−0.658***	−0.226	−0.080**	−0.043*	0.017
	(0.204)	(0.213)	(0.186)	(0.039)	(0.025)	(0.014)
CRI	0.008*	0.0002	0.002	0.0004	0.0001	−0.0004
	(0.004)	(0.004)	(0.004)	(0.001)	(0.0005)	(0.0003)
African dummy	−0.384	−0.655*	−0.681**	−0.059	−0.015	0.010
	(0.360)	(0.343)	(0.319)	(0.059)	(0.043)	(0.024)
LDCs	−0.780*	−1.089**	−0.171	0.065	−0.080	0.033
	(0.445)	(0.450)	(0.347)	(0.074)	(0.049)	(0.027)
SIDS	−0.666	−1.175*	0.682*	−0.059	0.002	−0.047
	(0.451)	(0.626)	(0.401)	(0.106)	(0.097)	(0.035)
ND-GAIN adapt. capacity	1.287	0.729	−2.175*	0.195	0.055	−0.110
	(1.316)	(1.374)	(1.134)	(0.237)	(0.179)	(0.089)
GDP per capita	1.242	2.597	−1.548	0.884***	0.380*	−0.061
	(1.769)	(1.917)	(1.592)	(0.305)	(0.227)	(0.133)
GDP per capita sq.	−0.131	−0.191	0.065	−0.056***	−0.026*	0.003
	(0.115)	(0.123)	(0.102)	(0.020)	(0.015)	(0.008)
WGI index	0.060	1.647***	0.946***	0.085	0.132***	0.022
	(0.302)	(0.329)	(0.262)	(0.054)	(0.037)	(0.019)

Exports	0.331**	−0.226**	0.112	0.004	−0.037**	−0.004
	(0.167)	(0.114)	(0.107)	(0.032)	(0.016)	(0.010)
Geographic distance	1.015***	0.381	1.009***	−0.103**	−0.004	0.017
	(0.237)	(0.258)	(0.288)	(0.044)	(0.030)	(0.024)
Colonial dummy			0.108			−0.044**
			(0.299)			(0.021)
UN joint voting	0.826	1.920**	−0.696	−0.026	0.083	0.0003
	(0.812)	(0.824)	(0.694)	(0.148)	(0.113)	(0.057)
Total development aid	0.627***	1.340***	0.363***	0.234***	0.114***	0.037***
	(0.116)	(0.127)	(0.098)	(0.021)	(0.013)	(0.008)
Population	0.024	0.494***	0.613***	−0.257***	−0.027	−0.025**
	(0.199)	(0.153)	(0.139)	(0.033)	(0.018)	(0.012)
Constant	−18.609**	−17.713**	−12.960*	1.195	−0.229	0.464
	(7.682)	(8.112)	(7.217)	(1.330)	(0.965)	(0.563)
Observations	720	720	720	463	245	280
R^2				0.460	0.414	0.400
Adjusted R^2				0.435	0.361	0.351
Log likelihood	−263.643	−250.772	−323.643			
Akaike inf. crit.	569.286	543.543	691.286			

Note: year fixed effects omitted
Standard errors in parentheses
$^{*}p < 0.1$; $^{**}p < 0.05$; $^{***}p < 0.01$

are significant at the selection stage, but only the EVI is significant at the allocation stage, and only when we also consider significant adaptation aid (model 4 in Table 5.6). Moreover, the EVI shows an unexpected negative sign, indicating that Germany is more likely to provide adaptation aid to less vulnerable countries and also provides less significant adaptation aid to vulnerable countries. African countries, LDCs, and SIDS do not receive more adaptation aid either. The effects are insignificant across the models (the effect for LDCs at the selection stage is weakly significant at the 90% confidence level) and partly have the wrong sign. Overall, the model provides no evidence that Germany prioritises physically vulnerable countries; if at all, Germany is more likely to support adaptation in *less* vulnerable countries.

We find slightly more evidence that Germany considers recipients' capacity to deal with the adverse effects of climate change. Particularly GDP per capita and its square term have significant coefficients with the expected sign in most of the German models. Similar to the aggregate case, poorer countries are much more likely to receive adaptation aid from Germany than wealthier countries, yet the actual numbers are much higher than in the aggregate case: the model for Germany predicts that 40% of the poorest countries receive adaptation aid, while almost none of the richest ones do so. These higher numbers reflect the higher amounts of adaptation aid provided by Germany as well as its involvement in a large number of countries. At the allocation stage, we also find similar patterns as in the aggregate models, with middle income countries expected to receive the highest levels of adaptation aid (up to about 35 cents per capita at a GDP per capita of around $3500). In addition, we see that the composite WGI index is significant at the selection stage, but not at the allocation stage (as in the aggregate models). In sum, adaptive capacity and good governance play some role for Germany's decisions on how to distribute adaptation aid.

What about donor interests? While trade ties presumably are particularly important for exporting nations such as Germany, we only find limited evidence that the country prioritises its trading partners. Only at the selection stage of the principal and discounted significant adaptation aid model do we find the expected positive and significant effect (model 1 in Table 5.6). The coefficient is insignificant in all other models. For the remaining donor interest models, the results are ambiguous. The colonial dummy is irrelevant for Germany. For geographic distance, countries that are further away are more likely to receive adaptation aid but receive less per capita. This effect is consistently significant in the models at the 95% or 99% confidence level. For voting similarity in the UN General Assembly, we similarly find that countries that vote in line with Germany are more likely to receive principal adaptation aid but then receive less per capita. This effect is significant for principal adaptation only; when we also consider significant adaptation aid, the significance of this effect disappears.

As is the case for all donors combined, German adaptation aid follows its development aid in general: countries that receive development aid from Germany are very likely to receive adaptation aid as well, and moreover tend to receive higher levels. Population also matters. As expected, large countries are more likely to receive some adaptation aid, but they receive less per capita.

Overall, the results for Germany are rather ambiguous. German adaptation aid does not seem to be strongly guided by either recipient need or donor interests. The country focuses its adaptation aid on poorer countries and well-governed countries, but also takes into account other factors—though not always in the way we would expect. The qualitative analysis in Chap. 6 sheds further light on these inconclusive findings and help us to understand better how Germany allocates its adaptation aid.

5.2.2 *Sweden*

Let us turn to the case of Sweden. Note that the results of the allocation stage—particularly for the principal adaptation aid only model—rest on a small number of cases, as Sweden provides less adaptation aid in total than for example Germany (see Chap. 4) and distributes this aid to fewer countries (see also Chap. 6).

At both aid allocation stages, there is only limited evidence that Sweden considers the physical vulnerability of recipient countries when deciding to whom to give adaptation aid, or how much to give to selected countries. The ND-GAIN exposure sub-index does not appear to play a role for Sweden either, as was the case for Germany, despite being a strong predictor of adaptation aid allocation in the general models. Also as in the German case, the EVI shows the unexpected negative coefficient and is significant across all models. The CRI is positive, as expected, but only significant at the selection stage of the principal adaptation aid only model (model 2 in Table 5.5). It should also be noted here that SIDS are not in the models of Table 5.5, since Sweden does not allocate any principal adaptation aid to countries in that group and estimation is therefore not possible. Surprisingly, we find a negative and mostly significant effect for

LDCs across the models, although Sweden explicitly seeks to support LDCs and very poor countries (see Chap. 6). The coefficient for African countries is also negative and weakly significant at the selection stage for principal and discounted significant adaptation aid (Table 5.6). Overall, like Germany, Sweden does not seem to prioritise vulnerable countries. If at all, the regression analysis indicates a negative relationship between vulnerability and Swedish adaptation aid.

Good governance, in contrast, is rather important for Sweden, which is also what our qualitative analysis suggests. Countries with high adaptive capacity and hence good governance as measured by the ND-GAIN adaptive capacity sub-index are more likely to receive principal adaptation aid (model 2 in Table 5.5). Leaving aside the allocation stage of Table 5.5 due to the small number of observations, we find a consistent positive and significant effect of the composite WGI index on adaptation aid: Sweden is more likely to support adaptation in well-governed countries and also tends to give more aid to these countries. Indeed, the model predicts that the worst governed countries have—all else being equal—an almost zero chance of receiving principal adaptation aid from Sweden, while it predicts a probability of about 74% of receiving principal adaptation aid for the best governed countries. Good governance seems to play an outstandingly large role in the case of Sweden. GDP per capita has a weakly significant and non-linear effect for the amount of principal and significant adaptation aid (model 4 in Table 5.6): countries at per capita incomes of around $1500 are expected to receive the highest levels of adaptation aid from Sweden, while richer countries (and some of the very poorest) receive significantly less.

Finally, political or economic ties are of only limited importance to Sweden for adaptation aid allocation. Indeed, countries to which Sweden exports more are *less* likely to receive Swedish adaptation aid, and also tend to receive less aid per capita, at least for principal and significant adaptation aid. Countries that are closer to Sweden are more likely to receive principal adaptation aid, while those that vote similarly to Sweden in the UN General Assembly are more likely to receive principal and significant adaptation aid. Overall, however, the results for donor interests are not very strong. Again, Swedish adaptation aid is closely linked to its development aid in general, and larger countries more often receive adaptation aid.

In sum, we find that Sweden focuses its adaptation aid on well-governed countries. The country does not seem to prioritise vulnerable countries: neither physically more exposed countries nor countries with low adaptive capacity receive more adaptation aid. Sweden is also not strongly guided by its own economic or political interests, which is in line with previous research that found Scandinavian countries to behave rather altruistically (e.g. Berthélemy 2006; Clist 2011) as well as with our qualitative analysis. Yet, Sweden also does not give more adaptation aid to poorer countries, even if previous research and our qualitative analysis suggest that the country has a strong focus on poor and least developed countries (see Chap. 6).

5.2.3 United Kingdom

Last but not least we take a look at British adaptation aid. The UK committed 'only' about as much principal adaptation aid as Sweden (both around $200 million since 2010, compared to almost $900 million for Germany for bilateral principal adaptation aid), yet it allocated this aid to a much wider range of recipients than Sweden. This can be seen by the much higher number of observations at the allocation stage of Table 5.5. This spreading out of limited resources across a large number of recipients probably explains why we have only very limited findings at the allocation stage. Only the control variables—total development aid and total population—are significant predictors at this stage for principal adaptation aid.

When we consider principal and discounted significant adaptation aid, Sweden and the UK not only distribute similar amounts of adaptation aid, but also target roughly similar numbers of recipients (Table 5.6)—though they seem to support different countries. In contrast to Germany and Sweden, but in line with the aggregate results, the UK takes into account physical vulnerability, as measured by the ND-GAIN physical exposure sub-index. Physically exposed countries not only are more likely to receive adaptation aid from the UK, but also tend to receive higher amounts per capita (at least when significant adaptation aid is also included). In contrast, neither the CRI nor the EVI influence how the UK distributes its adaptation aid. African countries are less likely and SIDS slightly more likely to receive British adaptation aid. Well-governed countries are also more likely to receive British adaptation aid, as both the negative and significant coefficient of the ND-GAIN adaptive capacity and the positive and significant coefficient of the composite WGI index suggest. GDP per capita does not play a role for the UK. Overall, this suggests that the

UK takes into account physical vulnerability as well as good governance, notably at the selection stage, but not adaptive capacity.

The models also do not find much evidence for donor interests. Contrary to our expectations, countries that are further away are *more* likely to receive adaptation aid from the UK. Former colonies receive less significant adaptation aid per capita, as we have already found for the aggregate case. Similarly, as for the aggregate analysis, and the analyses for Germany and Sweden, our control variables have a significant effect on British adaptation aid. Countries that receive British development aid also tend to receive its adaptation aid. Larger countries are more likely to receive adaptation aid but receive less per capita.

Overall, then, the UK focuses its adaptation aid on physically vulnerable countries, as measured by the ND-GAIN exposure sub-index, as well as on well-governed countries. Adaptive capacity or donor interests do not strongly influence how the UK distributes its adaptation aid.

This chapter has presented the results of our statistical analysis. We started by examining overall patterns in adaptation aid allocation across all OECD donors and found that recipient need, recipient merit, and donor interests (as well as our control variables of total development aid and population size) all influence how adaptation aid is distributed across recipient countries. We found rather strong evidence that physical vulnerability matters: countries that are more exposed and more sensitive to climate risks receive more adaptation aid, all else being equal. Adaptive capacity, in contrast, is a weaker predictor of adaptation aid, although poorer countries do receive more. Recipient merit, which is closely linked to adaptive capacity, also influences adaptation aid giving. Donors seem concerned with aid effectiveness and tend to give more adaptation aid to better governed countries which are (seen as) better able to use resources in an efficient and effective manner, even if these countries are also better able to cope with climate change and hence are less vulnerable. Finally, we found that donors take into account their own economic interests by allocating adaptation aid to countries to which they export a lot. In contrast, political interests do not seem to play a large role, maybe because we have chosen measures of political interests that are too broad to capture interests specific to the climate change arena. Political interests nonetheless are likely to matter, as indicated by the close link between total development aid and adaptation aid. Donors tend to support adaptation in those countries to which they provide development aid in general, and we know that development aid in general is distributed based on donor interests (see

Chap. 2). The close relationship between development aid and adaptation aid also puts into question the additionality of adaptation finance, a point to which we return in the concluding chapter.

Finally, we explored adaptation aid allocation in our three case study countries of Germany, Sweden, and the UK. This comparison hinted at considerable differences in how individual donors allocate their adaptation aid. Germany and Sweden did not seem to take into account physical vulnerability, which the UK did, at least according to one of our vulnerability measures, the ND-GAIN exposure sub-index. Good governance, while influencing all three donors, played a particularly important role for Sweden. The evidence for donor interests, in contrast, was rather weak for all three countries. The weak and overall rather inconclusive results for the three case study countries point to the need for qualitative in-depth analysis of adaptation aid allocation, to which we now turn in the next chapter.

Notes

1. It should be noted that when we compute predicted probabilities for former colonies, the lowest predicted probabilities of receiving adaptation aid are around 6% and the highest almost 17%. This shows how important it is to set the control variables appropriately.
2. Note that the number of observations in Tables 5.2 and 5.4 that present the results of the allocation stage is far lower than the number of observations in Tables 5.1 and 5.3 that present the results of the selection stage. In other words—and to reiterate a point previously made—we model the second stage only for those countries that receive some adaptation aid.
3. For principal and discounted significant adaptation aid, the expected values tend to be somewhat higher at the allocation stage, but not by much. The major difference of the second dependent variable is that the probabilities of being selected at the first stage are much higher.
4. With aid with significant adaptation objectives discounted at 50%.
5. While the coefficient may appear very small compared to, for example, the coefficient of the ND-GAIN exposure sub-index, this results from the different ranges of these variables. The CRI has a much wider range than the ND-GAIN, which leads to smaller coefficients. This in itself does not mean it is less important.
6. The sign of the coefficients, however, can vary, depending on how the variable in question is coded (i.e., depending on whether higher values indicate higher or lower adaptive capacity/good governance).
7. Note that the *x*-axis of the respective panel in Fig. 5.3 is on an exponential scale.

8. According to the organisation, the sub-index 'measure[s] to what extent a country is capable to minimize the adverse impact of climate change' (ND-GAIN 2013, 2).
9. The exact predicted likelihoods of receiving aid are 3.0% at no trade flows, 9.2% at $100,000, 12.0% at $1 million, 14.9% at $10 million, 18.3% at $100 million, 22.4% at $1 billion, 26.9% at $10 billion, and 32.1% at $100 billion.
10. Total development aid is the strongest predictor of adaptation aid flows, yet this is 'only' a control variable and we did not formulate a specific hypothesis relating to total development aid.

References

Alesina, A., and D. Dollar. 2000. Who Gives Foreign Aid to Whom and Why? *Journal of Economic Growth* 5: 33–63.

Barrett, S. 2014. Subnational Climate Justice? Adaptation Finance Distribution and Climate Vulnerability. *World Development* 58: 130–142.

Berthélemy, J.-C. 2006. Aid Allocation: Comparing Donors' Behaviours. *Swedish Economic Policy Review* 13: 75–109.

Betzold, C. 2015. Vulnerabilität, Demokratie, politische Interessen? Wie Deutschland seine Hilfe für Anpassung an den Klimawandel verteilt. [Vulnerability, Democracy, Political Interests? How Germany Allocates its Aid for Adaptation to Climate Change]. *Zeitschrift für Internationale Beziehungen* 22(1): 75–101.

Betzold, C., and F. Weiler. 2017. Allocation of Aid for Adaptation to Climate Change: Do Vulnerable Countries Receive More Support. *International Environmental Agreements: Politics, Law and Economics* 17(1): 17–36.

Clist, P. 2011. 25 Years of Aid Allocation Practice: Whither Selectivity? *World Development* 39(10): 1724–1734.

Dollar, D., and V. Levin. 2006. The Increasing Selectivity of Foreign Aid, 1984–2003. *World Development* 34(12): 2034–2046.

Dudley, L., and C. Montmarquette. 1976. A Model of the Supply of Bilateral Foreign Aid. *The American Economic Review* 66(1): 132–142.

Füssel, H.-M. 2009. Review and Quantitative Analysis of Indices of Climate Change Exposure, Adaptive Capacity, Sensitivity, and Impacts. Technical Report, Background Note to the World Development Report 2010.

Hadenius, A., and J. Teorell. 2007. Pathways from Authoritarianism. *Journal of Democracy* 18(1): 143–157.

Hoeffler, A., and V. Outram. 2011. Need, Merit, or Self-Interest—What Determines the Allocation of Aid? *Review of Development Economics* 15(2): 237–250.

Kaly, U., C. Pratt, and J. Mitchell. 2004. The Environmental Vulnerability Index (EVI) 2004. Secretariat of the Pacific Community Applied Geoscience and Technology Division (SOPAC) Technical Report 384. Available online at http://www.sopac.org/sopac/evi/Files/EVI%202004%20Technical%20Report.pdf.

Kreft, S., D. Eckstein, L. Dorsch, and L. Fischer. 2016. Global Climate Risk Index 2016. Who Suffers Most From Extreme Weather Events? Weather-Related Loss Events in 2014 and 1995 to 2014 . Available online at https://germanwatch.org/de/download/13503.pdf.

Lightfoot, S., G.A. Davies, and R. Johns. 2016. Needs and Interests: Understanding the British Public's Balancing of Aid Priorities. Paper presented at the Workshop 'The Domestic Dimensions of Development Cooperation', Institute for Development Policy and Management, 24–25 October, University of Antwerp, Antwerp, Belgium.

McKinlay, R.D. 1978. The German Aid Relationship: A Test of the Recipient Need and the Donor Interest Models of the Distribution of German Bilateral Aid 1961–70. *European Journal of Political Research* 6: 235–257.

Nakhooda, S., T. Fransen, T. Kuramochi, A. Caravani, A. Prizzon, N. Shimizu, H. Tilley, A. Halimanjaya, and B. Welham. 2013. Mobilising International Climate Finance: Lessons from the Fast-Start Finance Period. Institute for Global Environmental Strategies, World Resources Institute, and Overseas Development Institute.

ND-GAIN. 2013. University of Notre Dame Global Adaptation Index Detailed Methodology Report. Available online at http://www3.nd.edu/~nchawla/methodology.pdf. Last accessed 15 March 2017.

ND-GAIN. n.d. ND-GAIN: Notre Dame Global Adaptation Index. Available online at http://index.gain.org/. Last accessed 15 March 2017.

Neumayer, E. 2003. *The Pattern of Aid Giving: The Impact of Good Governance on Development Assistance*. London: Routledge.

Niang, I., O.C. Ruppel, M.A. Abdrabo, A. Essel, C. Lennard, J. Padgham, and P. Urquhart. 2014. Africa. In *Climate Change 2014: Impacts, Adaptation, and Vulnerability. Part A: Global and Sectoral Aspects. Contribution of Working Group II to the Fifth Assessment Report of the Intergovernmental Panel on Climate Change*, chap. 22, ed. C. Field, V. Barros, D. Dokken, K. Mach, M. Mastrandrea, T. Bilir, M. Chatterjee, K. Ebi, Y. Estrada, R. Genova, B. Girma, E. Kissel, A. Levy, S. MacCracken, P. Mastrandrea, and L. White, 1199–1265. Cambridge: Cambridge University Press.

Nurse, L.A., R.F. McLean, J. Agard, L.P. Briguglio, V. Duvat-Magnan, N. Pelesikoti, E. Tompkins, and A. Webb (2014). Small Islands. In *Climate Change 2014: Impacts, Adaptation, and Vulnerability. Part B: Regional Aspects. Contribution of Working Group II to the Fifth Assessment Report of the Intergovernmental Panel on Climate Change*, ed. V. Barros, C. Field, D. Dokken, M. Mastrandrea, K. Mach, T. Bilir, M. Chatterjee, K. Ebi, Y. Estrada, R. Genova, B. Girma, E. Kissel, A. Levy, S. MacCracken, P. Mastrandrea, and L. White, Chapter 29, 1613–1654. Cambridge: Cambridge University Press.

Remling, E., and Å. Persson. 2015. Who is Adaptation for? Vulnerability and Adaptation Benefits in Proposals Approved by the UNFCCC Adaptation Fund. *Climate and Development* 7(1): 16–34.

Robertsen, J., N. Francken, and N. Molenaers. 2015. Determinants Of The Flow Of Bilateral Adaptation-Related Climate Change Financing To Sub-Saharan African Countries. LICOS Discussion Paper 373/2015. Catholic University Leuven.

Robinson, S.-a., and M. Dornan. 2017. International Financing for Climate Change Adaptation in Small Island Developing States. *Regional Environmental Change* 17(4): 1103–1115. https://doi.org/10.1007/s10113-016-1085-1.

Stadelmann, M., Å. Persson, I. Ratajczak-Juszko, and A. Michaelowa. 2014. Equity and Cost-Effectiveness of Multilateral Adaptation Finance: Are They Friends or Foes? *International Environmental Agreements: Politics, Law and Economics* 14: 101–120.

Teorell, J., N. Charron, M. Samanni, S. Holmberg, and B. Rothstein. 2011. The Quality of Government Dataset, version 6Apr11. University of Gothenburg: The Quality of Government Institute, http://www.qog.pol.gu.se.

Voeten, E., A. Strezhnev, and M. Bailey. 2009. United Nations General Assembly Voting Data. Available online at https://dataverse.harvard.edu/dataset.xhtml?persistentId=hdl:1902.1/12379.

Weiler, F C. Klöck, and M. Dornan. 2017. Vulnerability or Donor Interests? The Allocation of Aid for Climate Change Adaptation. World Development. Forthcoming.

Younas, J. 2008. Motivations for Bilateral Aid Allocation: Altruism or Trade Benefits. *European Journal of Political Economy* 24: 661–674.

Zanger, S.C. 2000. Good Governance and European Aid: The Impact of Political Conditionality. *European Union Politics* 1(3): 293–317.

CHAPTER 6

Qualitative Analysis: Adaptation Aid in Germany, Sweden, and the United Kingdom

The previous two chapters provided an overview of who receives and who gives adaptation aid, and then analysed the most important drivers of adaptation aid allocation, first in a pooled sample of donor countries, and in a second step for the three selected case study countries. More specifically, Chap. 4 explored overall adaptation flows from 2010 through to 2015 and compared donors as well as recipients with regard to how much adaptation they gave or received. The data shows that adaptation aid was a small, but growing, share of development aid, with some donors investing more than 10% of their development aid in adaptation projects. Germany, Sweden, and the UK were among the largest adaptation donors, while SIDS were among the largest adaptation aid recipients, per capita as well as in percent of total development aid. Chapter 5 examined more systematically the drivers of adaptation aid. Our results suggest, among other things, that vulnerable countries indeed obtained relatively high levels of adaptation aid, at least when vulnerability is understood and measured as physical exposure and sensitivity to climate risks. Yet, there are important differences across donor countries. For instance, while we find that SIDS received significantly more adaptation aid than non-SIDS, not every donor gave more adaptation aid to the SIDS. Indeed, neither Germany nor Sweden supported SIDS specifically (see Chap. 5).

In this chapter—and building on the findings from previous ones—we look behind the numbers and seek to understand the decision-making process around (adaptation) aid allocation. As described in more detail in

C. Betzold and F. Weiler, *Development Aid and Adaptation to Climate Change in Developing Countries*,
https://doi.org/10.1007/978-3-319-64510-0_6

Chap. 3, we spoke with aid practitioners and observers in our three case study countries about the role of climate change and specifically adaptation in and for development cooperation as well as about the aid allocation process more broadly (for a list of interviews, see Chap. 3, and for the questions used, see Table A.2 in Appendix). Additionally, we reviewed key policy documents such as overall aid guidelines, coalition agreements, or the OECD peer review reports (see Chap. 3).

Based on these documents and our interviews, we first show that climate change and increasingly adaptation play a central role in development cooperation in all three case study countries. We assert that it is important to distinguish adaptation from development as well as from mitigation for equity and fairness reasons, though these distinctions are hard to make on the ground (Sect. 6.1). We then turn to the aid allocation process and investigate how the three determinants of aid allocation—recipient need or vulnerability, recipient merit, and donor interests—play out in this process (Sect. 6.2). The qualitative analysis confirms the results from the statistical analysis in Chap. 5, but sheds more light on the patterns we found. For example, the qualitative analysis highlights the fact that aid allocation is always a political decision. Donor interests are important in this decision, yet they comprise more dimensions than those included in the statistical analysis. Aid efficiency and effectiveness, the ability to mobilise additional resources (for instance from the private sector), or public visibility are factors that influence aid allocation decisions and that could be subsumed under a broad understanding of donor interests. Finally, we address other issues and challenges related to adaptation finance, notably questions of additionality, reporting and accounting, and the overall level of adaptation funding (Sect. 6.3). Interviewees thus emphasised that it is important to distinguish conceptually between development assistance and climate finance; that measuring and tracking adaptation aid flows is difficult because precise definitions and guidelines are lacking; and that current levels of funding are inadequate to cover the costs of dealing with climate change in developing countries.

6.1 Climate Change, Adaptation, and Development Cooperation

Before we delve further into the question of how recipient need, recipient merit, and donor interests drive adaptation aid allocation in Germany, Sweden, and the UK, let us first take a look at how important adaptation and adaptation aid are for these three donors. All three countries are large donors of both bilateral and multilateral aid, including in the area of climate change. Germany has not yet met the international goal of spending 0.7% of GNI; in 2013, its net ODA represented 0.38% of GNI, just below the OECD DAC average of 0.39% (OECD 2015, 101)—although both the previous and current governments confirmed their commitment to increasing the aid budget to meet the 0.7% target (Government of Germany 2009, 2013). Sweden and the UK, by contrast, have both met the 0.7% target. The UK met this target for the first time in 2013 and enshrined this target into law in 2015: the International Development (Official Development Assistance Target) Act 2015 requires that, from 2015 onwards, ODA amounts to 0.7% of GNI (Parliament of the United Kingdom 2015). Sweden 'has a long tradition of generous and ambitious aid' (Government of Sweden 2016, 4; see also e.g. Government of Sweden 2012, 2014a) and was the first country to meet the 0.7% target in 1975 (OECD 2013). Since 2006, Sweden has met its national target of providing 1% of GNI for aid (Government of Sweden 2014a; see also Government of Sweden 2013, 2014b; OECD 2013, 18).

6.1.1 Climate Change as a Priority Area

While only a small—but growing—share of total development aid targets climate change adaptation (see Chap. 4), there is no doubt that climate change is a political priority in Germany, Sweden, and the UK, including in development politics, as both interviews and policy documents emphasise.

Climate change, including adaptation, is a core priority for German politics and development cooperation. Germany is among the largest climate donors and 'has been consistently expanding its commitment in this area' (BMU and BMZ 2013, 4; BMZ 2013, 2015b; OECD 2015, 16). The coalition agreements of 2009 and of 2013 specifically commit Germany to assisting development countries deal with climate change impacts (Government of Germany 2009, 2013). The International Climate Initiative, established in 2008, is a key part of Germany's climate finance and its fast-start finance commitment; it received additional funding through the special Energy and Climate Fund (BMU 2013; BMZ 2013, 107). Furthermore, since 2011, all development projects have to be screened against environmental and climate concerns (DE1; DE4; BMZ 2013, 106f; OECD 2015, 40). Interviewees confirm that climate change and climate change adaptation are high on the German political agenda: 'adaptation is a topic that has spread with a lot of clout' says one interviewee (DE4), while another adds: 'climate politics is extremely important for German development cooperation. It is one of the big issues with very high priority' (DE3; see also e.g. DE5; DE8).

Sweden sees itself as a global leader on climate change mitigation and adaptation and on development (e.g. Government of Sweden 2011, 2014c, 2016). Climate change is a concern domestically as well as internationally (SW3; Government of Sweden 2011). Several government policy statements explicitly promise to increase Swedish climate and environmental aid as well as to climate-proof this aid (Government of Sweden 2010, 21, 2014c, 18, 2015, 11; see also Alliansen 2010, 35). Climate change is one of three thematic priorities in Swedish development cooperation, the other two being democracy and human rights as well as gender equality. These three priorities were selected as critical elements for poverty reduction and are to guide Swedish aid policy (Government of Sweden 2014a; OECD 2013, 45). Accordingly, 'environmental and climate aspects are a central basis for all development cooperation' (Government Offices of Sweden 2010). Between 2009 and 2012, Sweden had a Special Climate Change Initiative focused on adaptation (Wasielewski Ahlfors 2011) as part of its fast-start finance commitment, but turned to a mainstreaming approach thereafter (SW1; SW2; Dzebo and van Asselt 2014), though more needs to be done in this respect according to the OECD peer review (OECD 2013, 17, 45). One step toward more mainstreaming and a clear indicator of the importance given to climate issues was the government reshuffle in May 2016 that brought international climate issues and development cooperation together under the Minister for International Development Cooperation and Climate within the Ministry of Foreign Affairs[1] (see also SW5; SW4). The new 2016 Aid Policy Framework further promises that 'the government increases its focus on environment and climate change issues' (Government of Sweden 2016, 15).

In the UK, combating climate change is one of six priorities in the business plan 2011–2015 of the Department for International Development (DfID), which includes supporting adaptation as well as low-carbon growth in developing countries (DfID 2011). The 2010 Liberal Democrat Manifesto promised to 'ensure that the developing world is prepared to deal with the consequences of a changing climate' through additional finance (Liberal Democrats 2010), while the Conservative Party promised to 'work towards an ambitious global deal that will limit emissions and make available substantial financial resources for adaptation and mitigation' (Conservative Party 2010, 91; see also Conservative Party 2015). The 2010 coalition government set up the International Climate Fund to help 'the poorest people adapt to the effects of climate change on their lives and livelihoods' (DfID 2012, 19). Further, climate and environmental concerns are mainstreamed in British development cooperation (DfID 2012; OECD 2014).

In all three countries, as well as globally, the role of climate change, adaptation, and adaptation finance has increased over the past 10 years or so (e.g. DE12). In Sweden, interviewees note that climate change has received much attention since around 2007–2008, in line with the growing international focus on the topic, Al Gore's climate change film, and global climate negotiations leading to the 2009 Copenhagen Summit (SW6; SW5; SW1). In Germany, the last 5 years in particular have seen a rise in adaptation (DE3); by now, 'climate change and adaptation are the new buzzwords that everybody is talking about' (DE11). In the UK, the preferred term is 'resilience' rather than 'adaptation', because of its broader scope (UK3)—although in practice many people 'use resilience and adaptation quite interchangeably' (UK1).

6.1.2 *Adaptation or Development?*

That adaptation has taken centre stage in development cooperation is not surprising given the close links between adaptation and development, as we have already seen in Chap. 2 (see also e.g. SW4; DE6). Both interviews and policy documents understand climate change and sustainable development as linked, even inseparable; one without the other is impossible (e.g. DE1; Government of Sweden 2014a; Government Offices of Sweden 2010; Raabe et al. 2013). Aid practitioners recognise that climate change affects development and threatens past achievements:

> Our view generally is that climate change is a threat to our development cooperation, to put it broadly like that. It has the potential to slow down, or reverse, some of the gains that we have made (UK1; see also e.g. UK4; DE4; BMZ 2015b; DfID 2012).

Accordingly, development needs to be made climate-proof, 'prepared for whatever change that is going to come' (SW5; see also SW7; Government of Sweden 2010, 21). If development cooperation does not take into account climate change, it is a 'waste of money' (DE12). The British Conservative Party promised in 2010 they would 'work to make our aid "climate-smart"' (Conservative Party 2010, 118), and DfID seeks to become a 'climate SMART organisation' (DfID 2012, 18; emphasis in original). One observer uses the term 'climate compatible development', and adds that this 'resonates quite well in developing countries, because it has the focus on development' (UK5).

There is no doubt that making development climate-proof, climate-smart, or climate compatible is perfectly sensible and 'just constitutes good development practice' (UK1; see also e.g. DE9; DE10). But whether climate-proofing is adaptation is contested. While admitting to the difficulties of separating the two concepts on the ground, one interviewee explains that climate-proofing is a misnomer: 'Actually, it shouldn't be called climate-proofing. From a classical sustainability perspective, any [development] project should be designed so as to hold 50 to 100 years' (DE9). What, then, is adaptation? Although everybody talks about adaptation, there is no clear definition of adaptation. 'Climate adaptation could be almost anything', explains one interviewee (SW8). Another one similarly asks: 'How do you adapt to climate change? What does it mean?' (SW2); the answer, for another interviewee, is context dependent: 'exactly what is an adaptation intervention in a given context is really defined by that local context' (SW1). Adaptation is 'a very broad issue' (SW7)—hence the preference for the term 'resilience' over 'adaptation': 'we are rather taking about resilience, including climate adaptation. That is the trend' (SW2; see also UK3). Resilience also encompasses both adaptation to long-term climate *change* and shorter-term climate *variability*:

> Lots of people talk about adaptation when they actually mean building resilience to current variability. It is not in the strict definition of adaptation to long-term climate change. [...] That is why the terminology around resilience has come along (UK1; see also UK4 and Chap. 1).

The lack of clear definitions is problematic for reporting —and raises the question of additionality (see Sect. 6.3 below). Accordingly, an observer in Germany criticises the fact that many projects are now called adaptation projects, but are not necessarily directly about adaptation. Rural development, for instance, can contribute to climate change adaptation, but this does not happen automatically; projects need to take into account and reflect the risks from climate change (DE11; see also DE10). This stands in contrast to the comment by a government interviewee in the UK, who argues that

> there are a lot of things that are considered adaptation which aren't labelled as adaptation. Work on energy, work on innovation, work on strengthening basic services, they are not labelled as adaptation, although they could be (UK1).

From this perspective, adaptation is also nothing new. Particularly in Sweden, interviewees emphasised that much of their development work of the past 20 years or more is, at least in parts, work that is now labelled 'adaptation' (SW1; SW2; SW6; SW7). To some extent, then, the different labels—climate-proof, adaptation, resilience—are just 'artificial divisions of labelling' (UK1). On the other hand, observers also see a danger insofar as resilience, or climate-proofing, might supersede adaptation: 'I think personally that the move to resilience means that climate change adaptation might lose' (UK4; see also DE9; DE10).

6.1.3 Adaptation or Mitigation?

A related distinction is between adaptation and mitigation. Adaptation is not only related to development, but also to mitigation. While adaptation and mitigation are treated as separate issues in the negotiations (see Chap. 2), this has not always been the case. 'In the beginning, it was about climate. Really just about climate. It was only later that the distinction between adaptation and mitigation came up in the global debate' (DE4). The history of the Rio markers reflect this development; remember that the OECD first introduced a climate marker and only in 2009 decided to use two separate markers for adaptation and mitigation (OECD 2011). Indeed, mitigation and adaptation are linked; not only because there is 'no adaptation without mitigation', as one interviewee very concisely explains (DE1), but also because mitigation projects may have adaptation benefits and vice versa. In particular by working with an ecosystem approach, development cooperation kills two birds with one stone, as projects often 'have those benefits: it is good for adaptation and for mitigation' (SW1), and so the distinction is sometimes difficult to make, particularly on the ground (SW8; SW4; DE6).

Just as with the distinction between adaptation and development, however, the distinction between adaptation and mitigation is important from a justice perspective. The Copenhagen Accord and the Paris Agreement both call for a 'balanced' allocation of funds to adaptation and mitigation (UNFCCC 2009, Decision 2/CP.15, para. 8; UNFCCC 2015, Article 9, para. 4). Developing countries emphasise adaptation over mitigation (see Chap. 2). In practice, however, there is an imbalance between the two, with mitigation receiving much higher levels of funding.[2] Adaptation 'has traditionally been sidelined' (SW5; see also e.g. DE9). We see this imbalance also in our case study countries. In Germany, for instance, one interviewee admits that the country needs to focus more resources on adaptation in the future (DE3). Sweden, in contrast, has put more emphasis on adaptation, partly in response to other donors' focus on mitigation: 'Sweden also had the perception that because many other donors focus so much on mitigation, we had maybe an even stronger focus on adaptation', explains a government interviewee (SW1; see also SW8).

6.2 The Aid Allocation Process

Having discussed the importance of adaptation aid and its relationship to development and mitigation aid in our three countries, we turn to the core topic of this study, that is, how is adaptation aid allocated? Partly because adaptation is hard to distinguish from development more broadly, the decision-making process for allocating adaptation aid is closely linked to how development aid is allocated; after all, adaptation aid is a subset of overall development aid (see the discussion of adaptation finance versus adaptation aid in Chap. 2). This close link can also be seen in the statistical analyses—both in the pooled and the separate country models—presented in Chap. 5. Before examining the role of the three determinants in Germany, Sweden, and the UK in more detail, let us first briefly look at the framework of this overall aid allocation process.

There are at least three stages in any aid allocation decision. Donors have first to decide whether they want to give aid bilaterally or multilaterally. For bilateral aid, they then need to decide to which countries they want to give aid; and finally, which programmes and projects they want to support within the selected recipients.[3] All of these questions have an important political dimension. Bilateral aid, for instance, gives the donor more control as well as more visibility than multilateral aid (see Sect. 6.2.3 below and Chap. 4). Bilateral aid tends to focus on a small number of 'partner', 'focus', or 'priority' countries, and in line with global agreements on aid effectiveness (see OECD DAC 2005, 2008, 2011) and OECD recommendations donors seek to reduce the number of such partner countries on which they focus their aid (e.g. DE8; DE4). Accordingly, the UK now has 28 priority countries, down from 43 in 2010 (DfID 2015, 9; OECD 2014, 38). Germany currently has 50 partner countries, compared to 92 in 2005 and 57 in 2010 (OECD 2015, 46).[4] Sweden cut down the number of partner countries from 42 to 32 in 2013 (OECD 2013, 38f; see also SW1).[5] Additionally, all donors are also active in countries that are not focus countries, including through regional programmes and partners such as NGOs (e.g. DfID 2015, 9; BMZ 2013; SW7; SW8).

How these partner countries are selected is rather opaque, and politics clearly plays a role (e.g. UK1; UK4; DE7). Sweden explicitly describes its allocation decisions as political in nature: 'the decision as to which countries Swedish bilateral aid will focus on is a political decision' (Ministry for Foreign Affairs 2007, cited in OECD 2013, 41).[6] In contrast, both Germany and the UK officially list criteria for aid allocation. The 2009 coalition agreement in Germany thus explains: 'We will work with a limited number of partner countries. [...] Good governance, need, significance of our aid, security risks and strategic partnerships will be important aspects' (Government of Germany 2009, 128; see also BMZ 2013). The 2010 Bilateral Aid Review in the UK focused British aid on fewer countries, based on recipients' 'development need, the likely effectiveness of assistance and strategic fit with UK government priorities' (DfID 2011, 5; see also OECD 2014). Despite such lists of criteria, aid allocation remains a political process. Our interviewees confirmed that different factors such as poverty influence their countries allocation decisions, but that such decisions are never the outcome of a purely mechanical exercise. There is no formula which computes whether a country receives aid, and if so, how much, based on factors like income or governance (e.g. DE7; DE8).

Donors are in principle fairly free in terms of their selection of partner countries—that is, the geographic distribution of aid—although of course they take into account what other donors do (e.g. Davies and Klasen 2013; Government of Sweden 2016; Klasen and Davies 2011). In contrast, the selection of programmes and projects within recipients—that is, the thematic or sectoral distribution of aid—is made jointly with the recipients, as all donors emphasise. At least this stage of the allocation decision is not a top-down process but negotiated with the recipients. Usually, donor government and recipient government have regular dialogues or negotiations to develop country strategies and to identify priority areas in which to work together (e.g. SW1; DE4; DE1; DfID 2012; Government of Sweden 2016). And while in practice donors may approach a recipient and suggest work in a specific area, it is important that aid interventions formally must start with an 'impulse' from the recipient side, as German interview partners in particular highlight (DE4; DE1; DE8).

6.2.1 Recipient Need or Vulnerability to Climate Change

Need is clearly an important consideration for donors' aid allocation decisions, and need, in an adaptation context, comprises both poverty and vulnerability to climate change. Indeed, poverty and vulnerability are closely related, as policy documents acknowledge: 'it is the world's poorest people and the world's poorest countries—those who are also least responsible for environmental and climate change—that have most to lose from climate change' (DfID 2012, 2)—and who have the least capacity to cope with change (Government of Sweden 2014a; Government Offices of Sweden 2010; see also e.g. Raabe et al. 2013). Accordingly, all three countries emphasise in policy documents that they seek to focus their aid, including their adaptation aid, on poor, if not the poorest, countries, as well as fragile and conflict affected states, because of the greatest need in these countries (e.g. OECD 2013, 44, 2014, 2015; BMU 2013; Government of Germany 2009; Government of Sweden 2003, 2014a; Wasielewski Ahlfors 2011; DfID 2012, 19).

Interviewees confirmed that poverty as well as vulnerability play a role. 'Vulnerability is important, of course', says one government interviewee in Germany (DE3; see also DE1), and observers confirm this (e.g. DE11; DE10). In the UK, one observer similarly notes: 'I would say, by and large, the UK is quite good in prioritising poverty and vulnerability to climate change' (UK2; see also OECD 2014). Similarly, in Sweden, 'the overall notion is we should focus on the most vulnerable and the LDCs' (SW1; Government of Sweden 2016, 47f). This is also what we find in our quantitative analysis: by and large, poorer and more vulnerable countries are more likely to receive adaptation aid, and also tend to receive higher amounts of it, all else being equal (see Chap. 5).

But while poverty can be relatively easily measured, vulnerability to climate change impacts is neither clearly defined nor easily measured (see Chaps. 1 and 3). It is not always clear who is actually vulnerable, and how this affects, or should affect, funding, as one interviewee explains:

> I do think this questions of vulnerability, and who is vulnerable, is really challenging. Lots of countries are vulnerable, and some of these countries, such as Ethiopia and Bangladesh, are receiving a much higher proportion of finance flows than others. That's justified, they are vulnerable. But there are many vulnerable countries who are not receiving anywhere close to that scale. (UK6; see also e.g. DE10).

Vulnerability assessments and indicators exist and have been used, but are not without problems (DE3; see also DE10; Chap. 2). One solution is to follow global negotiations, which recognise the SIDS, the LDCs, and, to a lesser extent, the African countries as particularly vulnerable. There seems to be political will to become active in these countries (DE1; DE3)—after all, given their special status in the 1992 Convention, it is 'easy' to accord them special treatment (DE10). On the other hand, interviewees also recognise the limitations of these categories, as one observer comments:

> There are the LDCs, the SIDS. But of course there are also always borderline cases. Take the Philippines for example. It is neither an LDC nor a SIDS but of course environmental disasters threaten the Philippines every year. It does not belong to these two groups, but the Philippines is a very vulnerable state. So it's very difficult to say whether vulnerable countries receive the support they need (DE10).

Given the heterogeneity of the country groups—with South Africa for instance being part of the group of African countries, or Singapore a SIDS[7]—there is also opposition to this special treatment: 'A country like Honduras that is very strongly affected [by climate change] says of course that they find this [special treatment of SIDS] unfair', explains one interview partner (DE10).

We find in our quantitative analysis (Chap. 5) that vulnerable countries—vulnerable in terms of physical exposure and sensitivity, as measured by different indicators—tend to receive more support. Interviewees explain this relationship partly in terms of demand. Donors emphasise that they closely work with partners and jointly identify priority areas (see above). What partner countries want or demand influences what donors fund, and demand varies: 'the developing countries are divided, because they really have different needs. The developing countries are not a homogeneous group' (DE12). Some countries are more aware about climate change, often because they are more vulnerable (UK5) and hence more often ask specifically for adaptation projects: 'particularly vulnerable countries are of course particularly vocal in asking for support with adaptation' (DE12; see also DE5; DE11). For other countries, climate change is not so high on the agenda, for instance because other issues are more urgent, as in the countries of the Middle East and Northern Africa: 'there are some countries that say, yes, climate change is important, but right now, we have other problems' (DE4; see also DE3; DE5). Demand further depends on capacities—as well as the ease of accessing funds. While poorer countries are more interested in adaptation (compared to mitigation; DE3), (very) poor countries lack capacity. They are thus unable to articulate their adaptation needs (DE12), and ownership cannot be guaranteed (DE4; see also SW4). Accordingly, the level of adaptation aid may be lower despite high vulnerability. Finally, countries that struggle with attracting development funding use the 'climate card':

> Countries that otherwise would miss out on development aid are now interested in climate change. Out of self-interest, they pin their hopes on climate finance. You see this in Latin America. Latin America is fairly rich compared to Africa and no longer receives that much development aid, so they say we are particularly affected by climate change. [...] By now, almost all countries play the climate card. They recognise that it is almost impossible to obtain funding without [reference to] climate (DE3).

Vulnerability clearly plays a role in adaptation aid allocation. Donors explicitly seek to support the poorest and most vulnerable countries, those where the need is greatest. Policy documents as well as our interviews thus confirm the relationship between vulnerability and the level of adaptation aid we found in the quantitative analysis. Yet, we found this relationship only for the physical dimension of vulnerability in the statistical analysis; countries with low adaptive capacity in contrast received less adaptation aid according to our regression models. Our qualitative analysis sheds light on this finding. While countries that already feel the effects of climate change tend to be more aware and specifically ask for support with adaptation, countries that lack institutional and adaptive capacity are often unable to articulate this demand, and donors are reluctant to invest in countries where ownership cannot be guaranteed, despite potentially high levels of vulnerability in these countries.

6.2.2 *Recipient Merit*

That countries with low institutional and adaptive capacity receive less adaptation aid already points to the relevance of recipient merit. Interview partners and policy documents emphasise that good governance is clearly a criterion, even precondition, for working with a recipient country. Two additional aspects came up under the category of recipient merit: absorptive capacity on the one hand, and what we could term 'climate commitment' on the other, both of which are related to institutional capacity and thus to adaptive capacity. We discuss these three points in turn.

Good governance is an important factor for all three donors, since successful and sustainable development necessitates basic freedoms and human rights. Promoting good governance, the rule of law, and human rights are explicit goals of the German government (OECD 2015, 35). The German Charter for the Future accordingly states that 'human rights are central to sustainability in all its dimensions. [...] We are therefore determined to strengthen and deepen a human rights based approach across all of Germany's international development cooperation' (BMZ 2015a, 35f). The 2009 and 2013 coalition agreements similarly emphasise the importance of human rights and good governance as preconditions for development cooperation; only for humanitarian reasons will the German government cooperate with countries 'whose governance conflicts with our values' (Government of Germany 2013, 182; see also OECD 2015, 36; BMZ 2013, 158). In the UK, the government follows its 'golden thread' theory, 'which focuses on conditions that are enablers of development: the rule of law, the absence of conflict and corruption, property rights, and strong institutions' (OECD 2014, 36). Respecting human rights, governing well and transparently, fighting corruption, and being accountable to citizens are partnership principles; 'a deterioration in any of these areas can result in assistance being reduced, suspended or delayed' (OECD 2014, 68). Swedish development cooperation similarly pursues a rights based approach (Government of Sweden 2016). The overarching Policy for Global Development specifies that 'development cooperation will promote and be characterized by respect for human rights, democracy and good governance', among other things (Government of Sweden 2003, 59).[8] The yearly government policy statements highlight the Swedish government's work to advance respect for human rights and civil liberties globally, including through its development policy (e.g. Government of Sweden 2010, 24, 2012, 13, 2015, 11f).

Interviewees similarly emphasise that good governance is critical for the selection of partner countries: 'it is definitively a factor in the overall selection and cooperation with [recipient] governments, and which counties to act in' (SW1; see also e.g. DE11). Suspicions of mismanagement of funds and corruption might end cooperation. 'As soon as there are corruption issues, Sweden basically stops funding it', says one observer in Sweden (SW4), while another one comments for the UK: 'governance, ability to use money well, corruption, whichever indicator you want to use to build this, is part of making the decision on who should be given funding [...] and the lack of good governance can be the reason for not giving money'—even in cases that score high on the other criteria, that is, vulnerability to climate change and poverty (UK2). Zimbabwe is a case in point. While Zimbabwe is among the most vulnerable and poorest countries and therefore requires support with adaptation and development more broadly, corruption is widespread. Aid therefore is unlikely to reach those who need it the most: 'giving money to Mugabe does not mean giving money to people who need it. So, quite rightly, the UK does not give money to Mugabe' (UK2; see also DE5).

Good governance tends to correlate with institutional capacity, which in turn relates to absorptive capacity, or the capacity to use funds in an efficient and effective way. Absorptive capacity is a 'bottleneck', as interviewees emphasise:

> Poor countries first of all need a lot of capacity development before you can invest in them. If I went to a relatively poor country, say Burkina Faso, and I told them, you need a big project to promote drip irrigation and dams and what not, they would be overwhelmed (DE3; see also e.g. DE6; DE4).

Accordingly, capacity building may have to precede more thematic projects such as climate change adaptation projects. This may also explain why we find a negative effect for LDCs and African countries in our statistical analysis. According to our results, these two groups of countries are *less* likely to receive adaptation aid than other countries (see Chap. 5). Donors acknowledge the need for capacity building and enabling countries 'to utilise the means and opportunities made available to them' (Government Offices of Sweden 2010). A Swedish government officer for instance explains:

> our concern is also that the absorption capacity and the institutional capacity to plan and implement adaptation measures in many developing countries is the real bottleneck. So we do try to focus quite a bit of our efforts on capacity building (SW1).

In countries with a certain level of institutional capacity, efforts tend to lead to better results (DE4).

This aspect of achieving better results, in the context of adaptation aid, also has to do with what we term 'climate commitment'. Donors reward countries that are more committed to climate change issues and that they perceive as constructive partners in the climate change negotiations. This aspect also relates back to the question of ownership and demand, as countries where climate change is high on the agenda tend to ask specifically for support with climate change adaptation (see Sect. 6.2.1 above). Some countries are better at articulating these demands, for instance because they have specific climate change units in the government (DE4; DE12). It is easier to achieve ownership of climate change projects in such a context (DE5). From a donor perspective, working with committed countries is therefore attractive:

> The countries had to have shown clear interest in being pioneers of climate compatibility, taking it seriously. [...] Some countries happened to be more proactive than others. Ethiopia being a very strong partner, Kenya, Colombia. And in Asia, Bangladesh and Pakistan, interestingly. Peru, being the host of COP20, there was a lot of attention on climate issues at the time, and that has helped to push the agenda. Some countries are pioneers and we like to work with them, and their experience can be shared with others (UK5; see also DE5).

This climate commitment not only refers to domestic policies, but also engagement in the climate change negotiations, as one observer explains:

> In particular with regard to climate finance, Germany argues that they want to support countries that are themselves ambitious. A country that has very ambitious mitigation targets and that very ambitiously implements measures at home, and that additionally behaves in a constructive way at the international climate summits – they are the countries that will be prioritised when it comes to financial support (DE12).

6.2.3 Donor Interests

Our qualitative analysis as well as the quantitative analysis has shown that both recipient need and recipient merit influence donors' aid allocation decisions. What about donor interests? The quantitative analysis suggested that donors do take into account their own economic interests, and to a lesser extent their political interests (see Chap. 5), and our qualitative analysis confirms this. However, the qualitative analysis offers a broader understanding of donor interests that goes beyond trade or historical ties; donor interests also include aspects like efficiency and effectiveness or public visibility.

One observer comments for donors in general:

> Traditionally, politically, aid has been used as a political tool of foreign policy. As I said, the UK gives money to Commonwealth countries. The French give money to Francophone countries. The Belgians give money to their former colonies. The US gives the biggest amount of money to Israel. You know, Israel is by no means poor. They get aid because of political reasons. There is a political economy dimension to giving aid. It is another means of foreign policy. They give aid to countries where they want to influence something, they want to buy influence, or they want to maintain good relations. And so every developed country with a development budget uses that budget for their political purposes. There is politics behind that money, it is not just charity (UK2).

The UK acknowledges rather openly and explicitly that development cooperation is (also) about its own national interest:

> We believe that promoting global prosperity is both a moral duty and in our national interest. [...] Development represents tremendous value for money. In short, it is good for our economy, our safety, our health, and our future (DfID 2011, 2).

Value for money and aid effectiveness are crucial factors in British development cooperation, with results-based or evidence-based spending being keywords (UK1; DfID 2011, 1). DfID accordingly 'has a strong focus on results across all of its work and takes action to identify ways to improve implementation and ensure it delivers real and lasting results on the ground' (DfID 2015, 31; see also Conservative Party 2010; OECD 2014).

Politics and national interest are certainly also relevant for Germany. The 2009 coalition agreement emphasises German interests and German values: 'In pursuing the goals of our development policies, equal weight is given to our values and interests' (Government of Germany 2009, 127). The list of partner countries has been revised according to 'German interests (including strategic partnerships and global environmental goods)', among other things (BMZ 2013, 158). Our interview partners also highlight that interest influences aid allocation: 'Germany does not only decide where to give aid based on need, but also based on the general political climate', says one observer (DE12), while another adds: 'national interests, geostrategic interests [...] historical ties play a role. [...] Of course, that always plays a role' (DE8; see also DE5; DE7). This is not specific for adaptation aid, but applies to development aid more generally (DE4; DE1). While it is hard to assess to what extent interests matter, there is little doubt that countries where Germany hopes to 'get something out' are more likely to get development aid (DE12)—and this 'getting something out' could mean economic or political advantages, or simply international support:

> That Germany cooperates with so many countries has certainly also been motivated by foreign policy objectives, because supporting so many countries of course means that you also get a lot of support internationally (DE7).[9]

Compared to Germany and the UK, Sweden is probably the least guided by self-interest. In general, Swedish policy documents do not specifically refer to Swedish national interest as a factor for geographic aid allocation, as British and German policy documents do, although the 2016 Aid Policy Framework acknowledges that its 'development cooperation contributes to and establishes a basis for broader relations between Sweden and cooperation countries' (Government of Sweden 2016, 44). Scandinavian countries in general have a reputation of being not only very generous but also focused on need, and Sweden seems to be particularly altruistic. Even compared to Denmark and Norway, 'Sweden seemed to have the least national interest. But that doesn't have to be true, it might just be well hidden' (SW4). Another observer also acknowledges that national interests matter, but less so for Sweden compared to other countries: the role of donor interests 'also varies substantially between donor countries. Some countries are very strong [on self-interest ...]. You want to have something out of it. Sweden has been, I would say, quite altruistic' (SW6).

6.2.4 Additional Factors

While Sweden does not directly seem to be guided by its foreign policy interests, there are additional factors related to effectiveness and efficiency on the one hand, and influence and visibility on the other, that could be subsumed under donor interests broadly understood.

We have already seen the importance of effectiveness and efficiency for the UK, but these also matter for Sweden and Germany. All three donors seek to work in countries where their aid has the most impact, where they have a 'comparative advantage' (e.g. OECD 2014, 47; Government of Sweden 2016). One aspect in this context relates to path dependency: a history of working together increases the chances of future cooperation (UK2; DE1; Government Offices of Sweden 2010, 21). Existing working relations reduce transaction costs by building on existing structures, know-how, and expertise (DE7). In particular in the area of financial cooperation, having a local partner that has proven its ability to implement projects successfully may even be a precondition for some projects (DE3; see also SW2), or at least an indicator for the likely success of future projects. Donors are keen to keep working in countries where past projects have been implemented successfully: 'Maybe an implementation agency says that is a great country, they put our resources to very good use. So why should we leave a country in which things work well?' (DE7; see also SW1).

Working relations or even past project success is one aspect of effectiveness and efficiency; another aspect of these relates to the ability of a project to leverage additional funding. Remember that the industrialised countries promised to 'mobilise' $100 billion every year by 2020, from both public and private sources (see Chap. 2). In particular the UK seeks to bring in additional resources through its aid, as one observer explains:

> they want to have this more catalytic effect, they want that every dollar spent can generate ten other dollars of contributions from other countries. So that's the argument they want to find, that is why they always allocate when there are also other donors also allocating. It is important for them that there is additional funding to what they spend. That is the UK model (UK7).

There is a general trend to bring in private money: 'There is a huge interest in supporting private actors and also to help mobilise money from private actors. [...] The discourse is shifting, here like everywhere else, toward private finance and toward private actors' (SW4; see also UK3; DE6; BMZ

2011; DfID 2012; Government of Sweden 2016). While donors see the benefits of leveraging private finance, observers are critical of this approach because the private sector will only invest 'where there is some profit to be made' (DE6). Adaptation projects, however, tend to be non-profit. It is almost impossible to invest in adaptation and make a profit, so private finance 'is very, very difficult to get [...] for adaptation' (SW5; see also SW7; DE9). Additionally, investors require a certain level of development and institutional capacity, so countries like the LDCs with very low levels of development are unlikely to attract private finance, even with subsidies from donors (DE6). For these reasons, observers call for public resources for adaptation: 'we need public funds to safeguard that adaptation needs are met', says one interviewee (SW7), while another adds, 'we want the share of public finance to be as high as possible. And we want to go into countries that the private sector basically ignores' (DE6).

Finally, donors seek to use their resources in a way that maximises influence and, at least for Germany, visibility —which is why donors prefer bilateral over multilateral channels (see also Chap. 4). Bilateral aid gives donors much more control over where and for what their aid goes:

> If the money is going through their own aid agency, then they control it 100% in deciding where it goes, who it goes to, what it goes for. And more importantly, who does *not* get it. They can choose *not* to give it to people, or to give it to people. When they give it to the United Nations and multilateral banks, they have some control, but not total control (UK2; see also DE11; DE5).

Until 2013, Germany had an official target of spending one-third of its development budget multilaterally, and two-thirds bilaterally, 'so as to widen the scope of German development politics and to improve the effectiveness of used resources' (Government of Germany 2009, 129; OECD 2015, 36f). In practice, however, Germany tends to give only 10 to 15% of its funding for climate finance multilaterally (DE11; DE12). Sweden gives significant amounts of its development aid, including its adaptation aid, through multilateral channels (OECD 2013, 52; see Chap. 4). 'Multilateral cooperation constitutes a central channel for Swedish development cooperation in the environmental and climate area' and elsewhere (Government Offices of Sweden 2010, 21). Overall, bilateral aid flows nevertheless dominate, as they not only provide more control but are also more visible than multilateral contributions, which seems to be a factor for Germany.

Public announcements of finance get media attention, particular during large summits, and even more so during summits that take place in the donor country, such as the 2015 G7 summit (DE6) or the 2014 first Green Climate Fund donor summit (DE5) that both took place in Germany. In the UK, by contrast, visibility and public attention seems to be counter-productive given the criticism toward aid, and climate aid in particular, in some sections of the population:

> In the UK, a lot of our finance commitments are under the radar, at the moment at least. The minister doesn't really want to be associated with the new multibillion announcement on finance to Mozambique or whatever. Profile is not an overarching principle. [...] So when the UK makes climate announcements, they try not to draw too much attention (UK6).

6.3 Issues and Challenges

The previous section reviewed the three determinants of aid allocation: recipient need, that is, vulnerability to climate change, recipient merit, and donor interests. Our interviews confirm the results from our quantitative analysis: all three factors come into play, in different forms and to different degrees. Overall, however, aid allocation is a political decision. Many actors and factors are involved and there is no simple allocation formula. The qualitative analysis raised additional issues that our quantitative analysis is unable to detect, notably difficulties related to defining and measuring adaptation, as well as questions about the additionality of adaptation finance and related accounting and reporting problems (see also Chap. 2). We now discuss these issues in turn.

6.3.1 Additionality

We have already discussed that adaptation is a rather vague concept that is hard to distinguish from development, and that it is accordingly also difficult to distinguish adaptation aid from development assistance more generally (see Chap. 2). This debate has also come up very clearly in our interviews. While interviewees recognise that the distinction between adaptation and development is hard, if not impossible, to make on the ground, they also insist that it is important from a conceptual point of view: climate change is an additional burden for developing countries, which therefore require additional support. Climate change should not divert resources away from other development challenges: 'for us it is very important to make the distinction of the finance flows to ensure that development funding is not decreased, that is, to ensure that climate finance is additional to ODA', explains one observer (SW5), while another adds:

> Spendingwise, you can't have one project that is adaptation and one project that is ODA. It makes no sense. [...] The separation is relevant, but it is more relevant here in the policy accounting, [...] at the advocacy level. More as an argument to raise money (SW7; see also e.g. SW4; DE9; DE12; UK6).

In principle, donors accept this argument and have agreed to provide 'new and additional' resources for climate action in the developing world in the Copenhagen Accord—although the term 'new and additional' has disappeared from the Paris Agreement (see Chap. 2; see also e.g. SW5; SW4). The pledge of additional resources has been repeated domestically; the Liberal Democrats in the UK for instance promised that they would 'ensure that adaptation and mitigation measures are financed by industrialised nations on top of existing aid commitments' (Liberal Democrats 2010, 63), while the German Social Democrats similarly emphasise that climate-proof development 'must not come at the expense of efforts to reach the Millennium Development Goals. Funds for climate change mitigation and adaptation must therefore be additional to ODA commitments' (Raabe et al. 2013, 23; see also Olivier et al. 2013). But while politicians and parties understand the need for additionality and separating climate finance from development aid on paper, commitment to concrete action is lacking (DE12; UK2).

As we have seen previously, the discussion also revolves around the lack of a baseline. It is unclear as to which resources should be new and additional (see Chap. 2). Without an agreed baseline, the term 'new and additional' is just 'empty words' (DE12; see also e.g. SW5), open to interpretation. And different interpretations abound:

> [The donors] have different interpretations of that. Our interpretation is, it means over and above development assistance. Their interpretation is, it is just more development assistance. New and additional just simply means, this year's budget is new compared to last year's budget (UK2).

In Sweden, the debate in particular focused on 0.7% of GNI as a baseline. Given that Sweden has provided 1% of its GNI for development cooperation since 2006, in line with its domestic target, one interpretation is that the additional 0.3% that Sweden provides above the 0.7% target could count as 'new and additional' resources—which civil society strongly criticises (e.g. SW8; SW4; SW5).

6.3.2 *Reporting and Accounting*

Without clear baselines and agreed definitions, it is not only difficult to assess whether resources are new and additional, but also to monitor and track adaptation finance. Observers have been rather critical of the reporting and data reliability, including for the OECD CRS, since reporting guidelines are too vague and incoherent: 'In the end, every country reports in the way they see fit' (DE10). This entails a risk that funds are mislabelled as adaptation, that the adaptation relevance of projects is overstated, and that funds are double-counted against several objectives and pledges—all of which seems to take place, to differing degrees, and not necessarily on purpose. Projects that consider climate change in some way are readily counted as adaptation aid and reported as (additional) climate finance: 'they take the entire bilateral package [...] and relabel it as climate. But it does not contain as much climate as the label promises' (DE12; see also DE11; DE9). We discussed the question as to whether climate-proofing is adaptation earlier (Sect. 6.1.2), and whether adaptation aid counts as new and additional climate finance (Sect. 6.3.1). Donors on the one hand count regular development assistance as climate finance. On the other hand, they also double-count commitments against multiple pledges. A development project may have several environmental objectives at the same time, as well as addressing other objectives like health. While the multiplicity of objectives reflects the reality on the ground, namely that different areas are related (DE11), it also results in over-reporting and double-counting, as one interviewee comments: 'Sometimes you wonder where all these committed funds come from. But it is easy when you count it three or four times. Once as a health project, once as a water project, once as a climate project' (DE6; see also DE11).

How do donors, those who are responsible for reporting and accounting, see these issues of over-reporting, mislabelling, and double-counting? Donors are aware of the difficulties of reporting but take this task very seriously. As we have seen, there are no clear definitions of what adaptation is, and hence also no clear ways of measuring and reporting it. In practice, donors rely on OECD definitions and Rio markers, on which any additional guidelines are built (e.g. SW2; DE1). Yet, the Rio markers 'are also not very clear in the definitions, and that is a problem for all donors' (SW2; see also SW1; DE1). 'Applying the OECD markers [...] has its problem. It is not easy to apply the markers' (DE3). Part of the problem stems from who is applying the Rio markers, and how familiar they are with adaptation and the adaptation marker: 'There is a lot of room for interpretation and it depends a lot on the knowledge and capacity and competence of the person using the markers, what they actually signify', says one observer (SW5; see also SW4; DE9), while a government interviewee admits: 'depending on the knowledge you have you will mark different contributions differently' (SW2). Another data problem concerns projects with significant adaptation objectives, that is, projects where adaptation is a co-benefit. It may only be a small part of a given project that is about adaptation, but this would still justify applying the adaptation marker to the entire project, thus overstating the adaptation relevance and therefore the amount of adaptation aid (DE12). Sweden and Germany count 50% of significant projects as climate finance (DE1; DE3; SW5; see also AdaptationWatch (Weikmans et al.) 2016, 24), while the UK computes a percentage for each project depending on how much of the project targets adaptation:

> They are going through every single project and they are saying: is that relevant for climate change? Yes, building the bridge was, maybe you can classify that. And then they say what proportion of the project should we count (UK6; see also DE1; UK5).

Such difficulties notwithstanding, donors realise the importance of good quality data and take reporting seriously (e.g. DE2). One observer says for Finland: 'They actually think very hard what sort of score they assign to each project. They know that NGOs and others are watching. Nobody wants to be caught greenwashing' (UK5). Adaptation projects are really about adaptation, donors insist: 'if it says adaptation, it is adaptation' (DE1; see also DE3). When using the Rio markers, staff must make very clear why the project is about adaptation and what the causal mechanisms are (DE3; SW1). Even though studies find significant over-reporting in the OECD data (Adaptation Watch 2015; Donner et al. 2016; Junghans and Harmeling 2012; see also SW4), this finding may partly result from the methodology used: these studies have recoded projects listed in the OECD CRS based on project descriptions and found considerable discrepancies between the numerical Rio markers and the project descriptions. Yet, project descriptions in the OECD CRS are often filled in last, explains one government interviewee. Even when the project description does not suggest a connection to climate issues, this does not necessarily mean the project is not about climate change, so numerical codes like the Rio markers may actually be more robust than project descriptions (DE2). Still, the reporting system relies entirely on self-reporting and this is fundamentally problematic (UK2).

6.3.3 Level of Funding

Clearly, it is difficult to define and measure adaptation and accordingly to distinguish adaptation aid from development assistance, and the OECD data are by no means perfect, but arguably the most comprehensive and comparable data that is available (see Chap. 3). The OECD data suggested an overall increase in adaptation aid—though adaptation remains a rather small portion of total development aid (see Chap. 4). We conclude our qualitative analysis by reviewing our interviewees' comments on the level of funding and their expectations for the future.

In line with the growing attention paid to adaptation (see Sect. 6.1.1), the level of adaptation aid has been increasing. Donors emphasise their commitment to adaptation and climate more broadly. Sweden for instance is proud to be the largest per capita contributor to the Green Climate Fund (SW5; SW7; see also Government of Sweden 2015, 11). Sweden, together with Germany, also helps to keep the Adaptation Fund alive through regular contributions (DE10; SW7). German interview partners repeatedly mentioned Chancellor Angela Merkel's announcement to double climate finance from €2 billion to €4 billion by 2020 (DE4; DE5)[10]—note that the chancellor made this announcement during the 2015 G7 summit in Germany, which relates back to the visibility argument (see Sect. 6.2.4). Policy documents confirm the commitment to increase the overall development budget to reach the 0.7% target (Government of Germany 2009, 2013; CDU-CSU Fraktion im Bundestag 2013; Raabe et al. 2013), as does the 2015 International Development Act enshrining the 0.7% target in law in the UK (Parliament of the United Kingdom 2015).

Accordingly, further increases in adaptation aid are to be expected, not least after the Paris Agreement upgraded the status of adaptation with a separate global adaptation goal (e.g. DE4; SW7). To some extent, it is not so much political will that is the bottleneck of mobilising climate finance, but the capacity of both the donors and the recipients, as one aid agency staff explains: 'We could be doing much more if we had more staff and more resources. [...] We are actually at the limit of what we can do. So we cannot complain that we are provided with too little money for the implementation of adaptation [measures]' (DE3). Just pouring more money into adaptation, however, does not automatically lead to more, or more effective, adaptation:

> Of course we can see that there is a lot of need for adaptation finance. But that does not also mean that inserting more would automatically deliver more results. [...] Our concern is that the absorption capacity and the institutional capacity to plan and implement adaptation measures in many developing countries is the real bottleneck (SW1).

Whether the expected increases are effective and enough to meet developing countries' needs is questionable. Civil society calls for more ambitious targets and is rather pessimistic: 'I don't think there are grounds to be optimistic, that we will see the scale up of adaptation finance that we need. There was real reluctance in Paris to agree to a specific number on adaptation', concludes one interviewee (UK6; see also UK4). Another one similarly argues for more ambitious action: 'Donors just need to provide more funding for adaptation [...] Not enough is happening' (DE10; see also DE12; SW5; DE6). Not enough is happening in particular when you compare available funding to actual climate change impacts and adaptation needs: 'I think on the one hand that a lot is happening, and that is really positive. But on the other hand, whenever you work on climate change, you cannot avoid the feeling that it is never enough' (DE11).

This chapter has delved deeper into the aid allocation processes in Germany, Sweden, and the UK. We used semi-structured interviews and key policy documents to look 'behind the numbers' and to understand more deeply how donors make their aid allocation decisions, what role recipient need, recipient merit, and donor interests play in such decisions, and what additional factors or criteria are considered. We first gave an overview of the development aid architecture in the three selected countries and noted how climate change and adaptation are core priorities in development cooperation.

We then discussed the three determinants of aid allocation—recipient need, recipient merit, and donor interests—in detail. The qualitative analysis confirmed that these three factors play a role, if to differing degrees. Donors for example take into account recipients' vulnerability to climate change, but note how difficult it is to identify vulnerable countries. They also pointed to additional considerations such as the absorptive capacity of recipients—which may be at odds with vulnerability considerations: the poorest and least developed countries may need the most support but at the same time be the least able to put resources to good use. There are thus additional factors beyond recipient need, recipient merit, and donor interests that donors consider when allocating aid. In the end, aid allocation decisions are fundamentally political decisions.

Finally, we also addressed three points that the qualitative analysis brought up: first, the question of additionality. Adaptation finance must be additional to development aid for equity and fairness reasons, as donors acknowledge, but concrete action is often lacking. Second, the issue of reporting and accounting. Interviewees highlight that measuring and tracking adaptation aid flows is difficult without clear definitions, guidelines, and baselines. Finally, the level of funding for adaptation. Although there was some room for optimism and hope for increasing levels of adaptation finance in the future, interviewees also noted that the available funding is unlikely to meet needs.

Notes

1. The Minister for International Development Cooperation and Climate is at the same time Deputy Prime Minister, see http://www.government.se/press-releases/2016/05/government-reshuffle-25-may-2016/
2. Globally, the *Landscape of Climate Finance* reports estimate that about 17% of public climate finance focused on adaptation (Buchner et al. 2015, 9). In the OECD CRS, about 57% of all climate-relevant aid targets mitigation only, 25% targets adaptation only, and 18% targets adaptation as well as mitigation. See also Chap. 1.
3. Our quantitative analysis considers these three stages: we focus on bilateral aid (first stage) and separately model the selection (which countries) and allocation (which programmes and projects) stages.
4. Germany additionally works with 29 'cooperation countries' with 'focussed regional or thematic cooperation' (see BMZ 2013, 158f).

5. Partnerships were phased out in 12 of the 42 partner countries; Myanmar and South Sudan became new partner countries.
6. Interestingly, our quantitative analysis does not suggest that political and economic factors influence the allocation of Swedish adaptation aid. On the one hand, this may relate to the broader understanding of donor interests than what our variables capture. On the other hand, focusing for instance on LDCs and the poorest countries is also a political decision, but we would interpret this as recipient need in our quantitative analysis.
7. Singapore, like some other SIDS, is too wealthy to be eligible for ODA. This shows just how diverse the category of SIDS is.
8. The Policy further lists 'gender equality, the sustainable use of natural resources and protection of the environment, economic growth and social development and social security' (Government of Sweden 2003, 59).
9. The interview partner called this the 'water can principle' (*'Gießkannenprinzip'*), maybe best translated as 'shotgun approach' – that is, giving everybody a share of the pie.
10. For the full speech (in German), see https://www.g7germany.de/Content/DE/Rede/2015/05/2015-05-19-merkel-klimadialog.html

References

Adaptation Watch. 2015. Toward Mutual Accountability: The 2015 Adaptation Finance Transparency Gap Report. Policy Briefing. Available online at http://www.adaptationwatch.org/.

AdaptationWatch. (Weikmans et al.) 2016. Towards Transparency. The 2016 Adaptation Finance Transparency Gap Report. AdaptationWatch. White Paper. Available online at http://www.adaptationwatch.org/.

Alliansen. 2010, August. Jobbmanifestet: Alliansens Valmanifest 2010–2014. Stockholm.

BMU. 2013, September. The International Climate Initiative: Innovative Financing for Climate Protection. Federal Ministry for the Environment, Nature Conservation and Nuclear Safety, Berlin.

BMU and BMZ. 2013. Together for a Common Cause: Germany's Contribution to International Climate Financing. Bundesministerium für Umwelt, Naturschutz, Bau und Reaktorsicherheit and Bundesministerium für wirtschaftliche Zusammenarbeit und Entwicklung, Berlin.

BMZ. 2011, January. Klimawandel und Entwicklung. Information Brochure 1/2011 of the Federal Ministry for Economic Cooperation and Development, Berlin.

BMZ. 2013, May. 14. Entwicklungspolitischer Bericht der Bundesregierung. Weißbuch zur Entwicklungspolitik. Federal Ministry for Economic Cooperation and Development (BMZ), Berlin.

BMZ. 2015a. Charter for the Future: One World—Our Responsibility. Federal Ministry for Economic Cooperation and Development (BMZ), Bonn.

BMZ. 2015b, April. Climate Action in Practice. The Contribution of German Development Policy. Federal Ministry for Economic Cooperation and Development (BMZ), Berlin.

Buchner, B., C. Trabacchi, F. Mazza, D. Abramskiehn, and D. Wang. 2015, November. Global Landscape of Climate Finance 2015. Climate Policy Initiative, Venice.

CDU-CSU Fraktion im Bundestag. 2013, April. Entwicklungspolitisches Positionspapier. Arbeitsgruppe Wirtschaftliche Zusammenarbeit und Entwicklung, Berlin.

Conservative Party. 2010, April. Invitation to Join the Government of Britain: The Conservative Manifesto 2010. Conservative Campaign Headquarters, London.

Conservative Party. 2015. Strong Leadership: A Clear Economic Plan. A Brighter, More Secure Future. Conservative Campaign Headquarters, London.

Davies, R.B., and S. Klasen. 2013. Of Donor Coordination, Free-Riding, Darlings, and Orphans: The Dependence of Bilateral Aid Commitments on Other Bilateral Giving. CESifo Working Paper No. 4177.

DfID. 2011, May. Business Plan 2011–2015, Department for International Development. Department for International Development, London.

DfID. 2012, June. Supporting a Healthy Environment: A Fresh Approach to Our Work on the Environment. Department for International Development, London.

DfID. 2015, July. Annual Report and Accounts 2014–15. Department for International Development, London.

Donner, S.D., M. Kandlikar, and S. Webber. 2016. Measuring and tracking the flow of climate change adaptation aid to the developing world. *Environmental Research Letters* 11(5): 1–9.

Dzebo, A., and H. van Asselt. 2014. Swedish Development Cooperation and Climate Change: Is There Potential for Better Mainstreaming? Nordic Centre of Excellence for Strategic Adaptation Research NORD-STAR Policy Brief.

Government of Germany. 2009. Wachstum. Bildung. Zusammenhalt. Koalitionsvertrag zwischen CDU, CSU und FDP, 17. Legislaturperiode. Available online at https://www.bmi.bund.de/SharedDocs/Downloads/DE/Ministerium/koalitionsvertrag.pdf.

Government of Germany. 2013. Deutschlands Zukunft gestalten. Koalitionsvertrag zwischen CDU, CSU und SPD, 18. Legislaturperiode. Available online at https://www.bundesregierung.de/Content/DE/_Anlagen/2013/2013-12-17-koalitionsvertrag.pdf.

Government of Sweden. 2003, May. Shared Responsibility: Sweden's Policy for Global Development. Government Bill 2002/03:122, Stockholm.

Government of Sweden. 2010, October. Regeringsförklaringen 2010, Stockholm. Available online at http://www.regeringen.se/informationsmaterial/2010/10/regeringsforklaringen-5-oktober-2010/.

Government of Sweden. 2011, September. Regeringsförklaringen 2011, Stockholm. Available online at http://www.regeringen.se/informationsmaterial/2011/09/regeringsforklaringen-15-september-2011/.

Government of Sweden. 2012, September. Regeringsförklaringen 2012, Stockholm. Available online at http://www.regeringen.se/informationsmaterial/2012/09/regeringsforklaringen-18-september-2012/.

Government of Sweden. 2013, September. Regeringsförklaringen 2013, Stockholm. Available online at http://www.regeringen.se/informationsmaterial/2013/09/regeringsforklaringen-17-september-2013/.

Government of Sweden. 2014a, March. Aid Policy Framework—The Direction of Swedish Aid. Government Communication 2013/14:131.

Government of Sweden. 2014b. Överenskommelse mellan Socialdemokraterna och Miljöpartiet de gröna, Stockholm. Available online at http://www.socialdemokraterna.se/upload/Overenskommelse_mellan_Socialdemokraterna_och_Miljopartiet_de_grona.pdf.

Government of Sweden. 2014c, October. Regeringsförklaringen 2014, Stockholm. Available online at http://www.regeringen.se/informationsmaterial/2014/10/regeringsforklaringen-3-oktober-2014/.

Government of Sweden. 2015, September. Regeringsförklaringen 2015, Stockholm. Available online at http://www.regeringen.se/tal/2015/09/regeringsforklaringen-den-15-september-2015/.

Government of Sweden. 2016, December. Policyramverk för svenskt utvecklingssamarbete och humanitärt bistånd. Regeringens skrivelse 2016/17:60, Stockholm.

Government Offices of Sweden. 2010. Policy for Environmental and Climate Issues in Swedish Development Cooperation, 2010–2014. Ministry of Foreign Affairs Sweden, Stockholm.

Junghans, L., and S. Harmeling. 2012. Different Tales from Different Countries: A First Assessment of the OECD 'Adaptation Marker'. Germanwatch Briefing Paper.

Klasen, S., and R.B. Davies. 2011. Of Donor Coordination, Free-Riding, Darlings, and Orphans: The Dependence of Bilateral Aid Commitments on Other Bilateral Giving. Proceedings of the German Development Economics Conference, Berlin 2011, No. 47.

Liberal Democrats. 2010. Liberal Democrat Manifesto 2010. Liberal Democrats, London.

Ministry for Foreign Affairs. 2007, August. Guidelines: Decisive Factors for Country Focus. Memorandum of the Department for Development Policy of the Ministry for Foreign Affairs, Stockholm.

OECD. 2011. Handbook on the OECD-DAC Climate Markers. Available online at www.oecd.org/dac/stats/48785310.pdf.

OECD. 2013. oecd *Development Co-operation Peer Review: Sweden 2013*. Paris: OECD Publishing.

OECD. 2014. oecd *Development Co-operation Peer Reviews: United Kingdom 2014*. Paris: OECD Publishing.

OECD. 2015. oecd *Development Co-operation Peer Reviews: Germany 2015*. Paris: OECD Publishing.

OECD DAC. 2005. The Paris Declaration on Aid Effectiveness. Second High Level Forum on Joint Progress toward Enhanced Aid Effectiveness (Harmonisation, Alignment, and Results), Paris. Available online at http://www.oecd.org/dac/effectiveness/34428351.pdf.

OECD DAC. 2008. Accra Agenda for Agenda. Third High Level Forum on Aid Effectiveness, Accra. Available online at http://www.oecd.org/dac/effectiveness/thehighlevelforaonaideffectivenessahistory.htm.

OECD DAC. 2011. Busan Partnership for Effective Development Co-operation. Fourth High Level Forum on Aid Effectiveness, Busan. Available online at http://www.oecd.org/dac/effectiveness/thehighlevelforaonaideffectivenessahistory.htm.

Olivier, J., T. Leiter, and J. Linke. 2013, November. Adaptation Made to Measure. A Guidebook to the Design and Results-based Monitoring of Climate Change Adaptation Projects. Deutsche Gesellschaft für Internationale Zusammenarbeit (GIZ) GmbH, Eschborn.

Parliament of the United Kingdom. 2015. International Development (Official Development Assistance Target) Act 2015. Chapter 12. Available online at http://www.legislation.gov.uk/ukpga/2015/12/pdfs/ukpga_20150012_en.pdf.

Raabe, S., B. Kofler, B. Hendricks, S. Rebmann, and K. Roth. 2013. Eine Welt, eine Zukunft: Für eine nachhaltige und fortschrittliche sozialdemokratische Entwicklungspolitik. Arbeitsgruppe für wirtschaftliche Zusammenarbeit und Entwicklung der SPD-Bundestagsfraktion, Berlin.

UNFCCC. 2009. Report of the Conference of the Parties on its fifteenth session, held in Copenhagen from 7 to 19 December 2009. Addendum. Part Two: Action taken by the Conference of the Parties at its fifteenth session. Document FCCC/CP/2009/11/Add.1.

UNFCCC. 2015, December. Adoption of the Paris Agreement. Decision 1/CP.21. Document FCCC/CP/2015/10/Add.1.

Wasielewski Ahlfors, S. 2011. Adapting to a Changing Climate: The Swedish Government's Special Climate Change Initiative 2009–2012. Swedish International Development Agency (Sida), Stockholm.

CHAPTER 7

Conclusion

In this concluding chapter, we first briefly look back at the research question and the research design of this book, as well as the theoretical expectations we derived from theory and past work for our empirical analyses. We then draw together our findings from the empirical analyses, both quantitative and qualitative. We briefly summarise key results, with a focus on similarities and differences across the three stages of our empirical analysis. To what extent do donors keep their promises and provide adaptation aid with priority to the most vulnerable? What about other factors? How do recipient merit and donor interests influence adaptation aid allocation? How do these factors play out at the aggregate level for all donors combined, and at the country level for Germany, Sweden, and the UK? How do the results for the case study countries differ from the aggregate results, but also from each other? To what extent, and how, do the interviews confirm or depart from the statistical results?

Finally, we discuss the implications of our study for future research as well as for policy-making. We will consider the limitations of our research and suggest avenues for future research that builds on our study. For example we have only focused on the allocation of adaptation aid between *countries*; we have not paid attention to how adaptation aid is then distributed at the subnational level, and thus to the extent to which adaptation aid reaches the most vulnerable and needy populations within a recipient country. Similarly, we have not paid attention to how adaptation aid is used on the ground: what kind of projects are supported, and to

C. Betzold and F. Weiler, *Development Aid and Adaptation to Climate Change in Developing Countries*,
https://doi.org/10.1007/978-3-319-64510-0_7

what extent do the supported measures effectively reduce vulnerability and increase resilience?

Research on adaptation aid is still in its early stages, and there are many as yet unanswered questions. Our study takes one step toward a more comprehensive and detailed understanding of adaptation aid and its distribution. We hope that the book has shed light on the decision-making process of how adaptation aid is allocated, but also that our work contributes to a fruitful discussion among academics and practitioners about the role adaptation aid should play in the future, and how the funds should best be used.

7.1 Overview

Our research started out with the promise of industrialised countries to support adaptation to climate change in developing countries, especially in those 'particularly vulnerable' to climate change. While the 1992 Framework Convention already includes this pledge, the 2009 Copenhagen Accord turned it into concrete numbers: donors agreed to provide $30 billion between 2010 and 2012, and to mobilise $100 in 'new and additional' resources per year by 2020, for both mitigation and adaptation. To what extent do industrialised countries keep their promise? How much adaptation aid is available, where does this aid come from, and how is it allocated to the recipient countries (see e.g. Peterson Carvalho and Terpstra 2015)? Do donors really 'tak[e] into account the urgent and immediate needs of developing countries that are particularly vulnerable to the adverse impacts of climate change' (UNFCCC 2007, e.g. Decision 1/CP.13, para. 1(c)), as they have promised repeatedly during the UNFCCC climate change negotiations (see Chap. 2)? To answer these questions, we focus on bilateral ODA targeting climate change adaptation in recipient countries—what we call adaptation aid throughout the book. While adaptation aid and adaptation finance are conceptually different (Bird 2011), donors do draw on their aid budgets to fulfil adaptation finance commitments: most adaptation finance to date has come as adaptation aid (e.g. Ayers and Abeysinghe 2013; Weikmans 2016). There are good reasons to support adaptation through development aid, but if donors do so, they must respect their international commitments, notably the focus on vulnerable countries.

Drawing on the large literature on development aid as well as the much smaller but growing literature on adaptation aid and the policy discussions just mentioned, we derived three major expectations for adaptation aid allocation (see Chap. 2). Our first hypothesis (H1a and H1b) builds on the recipient need model of aid allocation and expects donors to be guided by recipient need, or vulnerability to climate change impacts. In other words, more vulnerable countries should receive more adaptation aid—as donors promised. Having discussed vulnerability at some length, we divided this expectation into two parts, in line with the two dimensions of vulnerability (see particularly Sect. 1.1 in Chap. 1). First, vulnerability has a physical dimension, namely the degree to which countries are physically affected by climate change, that is, their exposure and sensitivity to climate risks. The more physically exposed and sensitive countries are, the stronger the impact a warming climate has on them, and the more adaptation aid they should receive (H1a). Second, vulnerability also has a socio-economic dimension, namely the degree to which countries are able to cope with, recover from, and adapt to climate change impacts, that is, their adaptive capacity. The lower a country's adaptive capacity is, the more vulnerable it is, and the more adaptation aid it should receive (H1b).

In contrast to the recipient need model of aid allocation, the donor interest model of aid allocation posits that donors are mainly motivated by their own self-interests: donors use aid, including adaptation aid, to promote their own foreign policy objectives. Accordingly, recipients should receive higher amounts of adaptation aid the more important they are for a specific donor—from an economic, political, or security point of view. Our hypothesis H3 accordingly expects that donors give more adaptation aid to economically or politically relevant countries. This may imply that donors do not, or only partially, keep their promises to direct adaptation funds to the countries most in need, if we assume that the most vulnerable countries are not necessarily the most economically or politically relevant countries.

A third model of aid allocation is more recent: the recipient merit model. This model argues that donors 'reward' countries that are well governed and implement the 'right' policies by allocating more aid to such countries, for two reasons. First, better governed countries use resources in a more efficient way; allocating aid to such countries hence increases aid effectiveness. Second, donors intrinsically value and seek to promote good economic and political governance. In other words, when donors consider recipient merit, they both make sure that their aid—that is, taxpayers' money—is used efficiently, and at the same time that they reward and promote democratic institutions. Our hypothesis H2 accordingly expects well governed countries to receive more adaptation aid. As we discuss, good governance is not always clearly separable from recipient need because of the close links between good governance and adaptive capacity.

To test these three hypotheses on aid allocation decisions across all donors, we compiled a dataset using data from the OECD CRS (see OECD 2016) and the Rio marker for adaptation to construct measures of adaptation aid flows (see Chap. 3, especially Sect. 3.1, including for limitations of this data). For all recipient–donor dyads and each year from 2010 through to 2015, we obtained a record of adaptation aid flows, and then calculated (a) the amount of principal adaptation aid per capita from donor j to recipient i in each year t; and (b) the amount of principal and discounted (at 50%) significant adaptation aid per capita from donor j to recipient i in each year t. We used these two dependent variables in the descriptive analysis in Chap. 4 and notably in the statistical models of Chap. 5. Table 3.1 in Chap. 3 lists all independent variables used to operationalise the three hypotheses, and Table A.1 in the Appendix provides summary statistics for all variables.

To investigate adaptation aid allocation decisions in more detail, we took a closer look at three carefully selected donor countries: Germany, Sweden, and the UK (see Sect. 3.3 in Chap. 3). In a first step, we repeated the quantitative analysis for these three countries and tested the three aid allocation models separately for them. This allowed us to draw conclusions about how our three country cases relate to the overall sample of all donor countries, but also to compare directly the results of the three country models. In a further step, we investigated the underlying decision-making processes in Germany, Sweden, and the UK (see Chap. 6), based on semi-structured interviews with policy-makers, aid practitioners, and experts in the three countries as well as key policy documents. Table 3.2 lists all interviews and institutions who participated in our research endeavour; Table 3.3 shows all policy documents used in the analysis; and Table A.2 in the Appendix describes the questionnaire that guided our semi-structured interviews.

7.2 Results

Over the time period under investigation in this study—the 6 years from 2010 through to 2015—donor countries committed just under $75 billion for adaptation to recipient countries, or approximately 4.8% of total development aid provided over that same period of time. The majority of these adaptation aid commitments—$50.9 billion—had 'only' significant adaptation objectives, while the remaining projects worth $23.9 billion had principal adaptation aid objectives. Almost 72% of all adaptation aid flows with either significant or principal adaptation objectives were provided through bilateral channels, while around 28% were distributed via multilateral funds and institutions (see Sect. 4.1 in Chap. 4). Over the years, adaptation aid has slowly been growing in absolute terms and as a share of total development aid. While it only made up about 3.8% of all development aid in 2010, the first year donors reported adaptation aid flows to the OECD; this fraction grew to 6.3% in 2015. This increase demonstrates that adaptation aid is an increasingly prominent form of development assistance, and we expect its significance to further increase in the years to come. It also indicates that developed countries—on aggregate—are serious about their promise to scale up climate and adaptation finance, including through development aid.

All three selected countries—Germany, Sweden, and the UK—are among the six largest donors. Germany has so far committed $11.9 billion for significant and principal adaptation projects, second only to Japan ($13.5 billion). Note though that Germany has disbursed only about half that sum ($6.2 billion) in the period of analysis. For commitments, the UK and Sweden are in fourth and sixth place, respectively, with $4.4 and $3.2 billion committed to adaptation projects, with disbursements in excess of (UK) or equal to (Sweden) commitments. In terms of disbursements, the UK was the largest adaptation donor, followed by Germany, Japan,[1] and Sweden. Sweden was also the single largest donor of adaption aid on a per capita basis, in terms of both committed and disbursed adaptation aid (closely followed by Norway and Denmark, see also Fig. 4.5 in Chap. 4).

The largest recipients of adaptation aid are mostly populous countries in Asia such as Vietnam, the Philippines, Bangladesh, or Indonesia—countries that are also known as fairly vulnerable to the impacts of climate change. On a per capita basis, each inhabitant of the recipient countries in our dataset combined received on average about $7.20 in significant and principal adaptation aid over the 6 years covered by our study. However, this ranges from only a couple of cents in countries such as Kazakhstan, Argentina, or China, to thousands of dollars in some very small SIDS such as Niue and Tuvalu. At least at first glance, donors seem to consider the vulnerability (in terms of physical exposure and sensitivity) of recipient countries when making adaptation aid allocation decisions.

Turning to our statistical analysis in Chap. 5, we first of all find very similar results (and hence allocation patterns) for both dependent variables under investigation—principal per capita adaptation aid on the one hand, and principal plus discounted significant per capita adaptation aid on the other. The results reported in Tables 5.1, 5.2, 5.3, and 5.4 are also very stable across the partial models. In other words, the three partial models on physical vulnerability, adaptive capacity and recipient merit, and donor interests are rather similar to the full model. Thus, our statistical results seem to be fairly reliable, for both the selection stage, at which donors decide to which countries they provide adaptation aid, and the allocation stage, at which donors decide how much adaptation aid they allocate to countries selected at the first stage. Figure 7.1 summarises the statistically significant results of the full models for principal adaptation aid only, for the selection stage and the allocation stage, shown in Tables 5.1 and 5.2 in Chap. 5. In the figure we juxtapose, for both stages, the *size* of the effects of

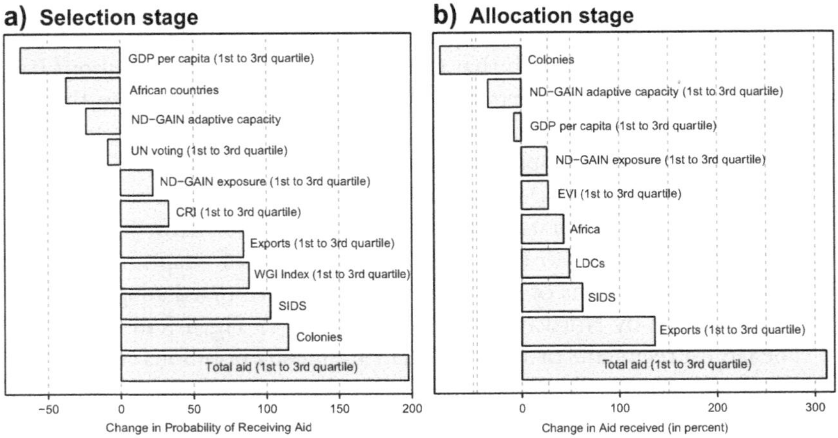

Fig. 7.1 All substantive effects for the full model for both adaptation aid allocation stages. (**a**) Selection stage. (**b**) Allocation stage

the various indicators we used to capture our hypotheses. More specifically, we show at the selection stage the *change in the probability* of being selected—all else being equal—when the variable in question changes (from the first to the third quartile for numerical variables, from zero to one for our dummies). At the allocation stage we show the *percentage change* in adaptation aid received for all the significant variables in the model.

As Fig. 7.1 shows, we find evidence in support of all three hypotheses: recipient need, recipient merit, and donor interests all help us to understand how adaptation aid is distributed, for both allocation stages. We have three measures of physical vulnerability, one of which—the ND-GAIN exposure sub-index—is significant at both stages of all models. One measure—the CRI—provides additional evidence for physical exposure playing a role at the selection stage, as does our third measure of physical vulnerability—the EVI—at the allocation stage. Taken together, the results for physical vulnerability indicate that donors allocated more adaptation aid to countries that are exposed and sensitive to climate risks, as expected (H1a).

We also find some evidence that adaptive capacity matters (H1b), though the evidence is less strong than for physical vulnerability. The results for GDP per capita shows that the richer recipients are, the less likely they are to receive some adaptation aid, and the less funding they receive in the case where they did receive some adaptation aid. The results of our other measures of adaptive capacity, in contrast, suggest that governance is more important than adaptive capacity, as our second hypothesis (H2) predicts. The results of the ND-GAIN adaptive capacity sub-index and the composite WGI index indicate that well-governed countries receive more adaptation aid, presumably because donors consider them to be able to use funding more effectively, even though such countries can also be seen as better able to deal with the challenges of climate change and hence are less vulnerable.

Finally, using trade flows from donor to recipient countries, we find strong evidence in favour of the donor interest model of aid allocation (H3). Developed countries use adaptation aid to promote their own economic interests by allocating more funds to their trading partners. Political interests, on the other hand, do not appear to be very important for adaptation aid allocation decisions. We find no or only a very weak relationship between adaptation aid and geographic distance, former colonial ties, or voting behaviour in the UN General Assembly. Given that adaptation aid is closely linked to development aid more broadly, however, we should not conclude that donor interests do not matter at all. Rather, our measures of foreign policy interests may be too broad to capture donors' *climate* interests, as our qualitative analysis also indicates.

When we take a closer look at the individual results for Germany, Sweden, and the UK, we find that the various hypotheses are supported by one country case or the other, at either the selection or the allocation stage. Overall, however, the results are much weaker than those of the aggregate analysis across all donors, partly because of the smaller number of observations. Adaptation aid allocation patterns in the three selected countries seem to be quite divergent. This should not be too surprising, especially as we selected our case study countries based on a most different systems design. While the UK for instance did consider physical vulnerability according to the statistical results, we did not find such an effect for Germany and Sweden. Sweden, in contrast, paid particular attention to good governance. Political interests did not seem to affect adaptation aid allocation in any of the three countries.

Our qualitative in-depth analysis based on interviews and policy documents helped to shed more light on these findings (see Chap. 6). The interviews and policy documents strongly suggest that donors take into account recipient need—both in terms of physical vulnerability and adaptive capacity. Poverty is an important criterion for allocating adaptation aid in all three countries. Interviewees and policy documents emphasise that donors seek to support adaptation in vulnerable, poor, and least developed countries in particular. This commitment is in line with the findings of the quantitative analysis across all donors. However, for the UK, we do not find evidence in the statistical analysis that poverty is indeed a criterion for how British adaptation aid is distributed, while we find that Germany and Sweden consider poverty but not as strongly as the aggregate results would have suggested. Similarly, there is strong statistical evidence that donors—in general—consider physical exposure and sensitivity in their decision-making processes, but only limited evidence that Sweden or Germany do so. These diverging results demonstrate the value of looking at both patterns in adaptation aid allocation in aggregated form and trends for individual countries. They also underline the need for qualitative analyses that 'look behind the numbers'.

We find similar patterns for the other two hypotheses, which are supported by the general models, but only partially by individual country data. Interviews and policy documents suggest that good governance plays a critical role for donors' decisions to provide adaptation aid. However, the interviews also suggest that for very poor countries institutional capacity often needs to be built before large-scale adaptation projects can be initialised. This goes back to the discussion of how recipient need (adaptive capacity) and recipient merit (good governance) are related and cannot easily be disentangled, which was problematic particularly for the quantitative analysis. If building institutions and capacity is a precursor of adaptation aid, should funds targeting such developments then also be considered adaptation aid? The case studies suggest that donors implicitly incorporate this connection in their adaptation aid allocation decisions, while the statistical models have a much harder time to pick up such intricacies.

Finally, the qualitative analysis also supports the findings that donor interests are—to various degrees—at play when adaptation aid is distributed. The UK acknowledges rather explicitly that development cooperation in general is in the country's interest and hence a political tool of foreign and economic policy. The interviews in Germany similarly show that political considerations motivate the distribution of adaptation aid. Other than the statistical analysis suggests, political motivations therefore seem to trump economic ones—at least in Germany and the UK. In Sweden, on the other hand, neither the qualitative nor the quantitative analysis point to a strong role of self-interest in development cooperation, including in adaptation aid allocation, although the country explicitly describes its allocation decision as 'political' in nature (Ministry for Foreign Affairs 2007, cited in OECD 2013, 41). The qualitative analysis further confirms the strong link between adaptation aid and development aid. There is no separate decision-making process for allocating adaptation aid. Donors hence support adaptation in countries in which they have been engaged in development cooperation in the past. This puts into question the additionality of adaptation aid to regular development aid, a point that was raised repeatedly in interviews and that we will address in the next section.

7.3 Implications

Having discussed the methods and results of our analyses, we now briefly turn to the empirical and theoretical implications of our study, and finally discuss the policy implications of our findings.

Empirically, our analysis provides the most comprehensive assessment of adaptation aid allocation to date. Using quantitative and qualitative methods, we show that the three models of aid allocation—recipient need, recipient merit, and donor interests—also influence the distribution of adaptation aid (see Fig. 7.1 and summary of results above). Theoretically, our study complements the broader aid allocation literature. We show that the three basic models of aid allocation apply to different types of aid. Yet while recipient merit and donor interests remain stable across different types of aid, we show that recipient need is a concept that is broader than just poverty and specific to the type of aid studied. In the context of climate change adaptation, recipient need translates into vulnerability to climate risks. The recipient model of aid allocation thus predicts that countries more vulnerable to climate change should receive more adaptation aid. As we discussed at greater length in Chaps. 1 and 2, identifying vulnerable countries is, however, a difficult task and necessarily involves political decisions (e.g. Klein 2009).

This brings us to the policy implications. We are specifically interested in whether donor countries keep what they promised during the international climate change negotiations, namely to assist 'particularly vulnerable' developing countries deal with the effects of climate change (see Chap. 2). While the 1992 Convention only vaguely stipulates that developed countries assist 'particularly vulnerable' developing countries 'in meeting costs of adaptation to those adverse effects [of climate change]' (UNFCCC 1992, Article 4.4), the 2009 Copenhagen Accord requires 'scaled up, new and additional, predictable and adequate funding' for climate change mitigation and adaptation, with adaptation and funding for adaptation to focus on 'reducing vulnerability and building resilience in developing countries, especially in those that are particularly vulnerable, especially least developed countries, small island developing States and Africa' (UNFCCC 2009, Decision 2/CP.15, para. 3 and 8). Donors hence formally agreed to allocate their support for adaptation based on recipient need. Donors further agreed to provide this support for adaptation *additionally* to other forms of development assistance.

The 'new and additional' aspect of adaptation finance has since disappeared from newer UNFCCC documents, notably the Paris Agreement. While the Paris Agreement does call on developed countries to 'significantly increas[e] adaptation finance from current levels and to further provide appropriate technology and capacity-building support' (UNFCCC 2015, Preamble, para. 114), it no longer specifies that such finance be new and additional, nor does it include any concrete financial commitment (see also e.g. Roberts and Weikmans 2015). Not surprisingly, then, there is some concern that donor countries are—at least to some degree—simply rebranding development aid as adaptation (or mitigation) finance (see e.g. Ayers and Abeysinghe 2013; Carty et al. 2016; Ciplet et al. 2013; Roberts and Weikmans 2017; see also Chap. 6). As there is no clear and agreed baseline for what constitutes 'new and additional' adaptation aid, such claims are hard to assess (Roberts and Weikmans 2017; Stadelmann et al. 2010). Additionally, there are clear synergies between adaptation and development, which not only makes it difficult to separate the two forms of assistance, but also risks duplication of efforts and misallocation of funds (see Smith et al. 2011). According to donors' own classification, adaptation aid has grown in importance, from $8.9 billion in 2010 to $20.3 billion in 2015. This is by no means enough to meet the global climate challenge (see Ha et al. 2016) and moreover is probably an optimistic upper bound of actual adaptation aid flows, given problems of over-reporting in the OECD data (see e.g. AdaptationWatch (Weikmans et al.) 2016; Donner et al. 2016; Junghans and Harmeling 2012; Roberts and Weikmans 2017).

Given these problems, it is hard to assess to what extent donors have kept their promise of providing or 'mobilising' new and additional funds. It is also hard to assess to what extent donors have kept their promise of prioritising 'particularly vulnerable' countries because identifying vulnerable countries is partly a political decision (Klein 2009; Roberts et al. 2017). Therefore, we left out these questions and instead examined the development assistance that donors themselves marked as relevant for adaptation, regardless of whether these resources were new and additional, and related this adaptation aid to various measures of physical vulnerability and adaptive capacity. Even if we leave aside the question of what baseline to use for identifying additional resources and the question of how reliably quantitative indicators can measure vulnerability, our analysis addresses both these questions. Indeed, allocation and additionality are related: if adaptation aid was new and additional to, and hence different from, 'regular' development aid, the two forms of assistance should be distributed according to different logics. Adaptation aid should mainly flow to vulnerable countries and not to the same countries that receive 'normal' development aid.

In practice, however, we find that total development aid is by far the strongest driver of adaptation aid allocation. At both the selection stage, and in particular at the allocation stage (see Fig. 7.1 above), we find a strong effect of total development aid. This means that funding for adaptation to a large degree follows development aid: donors provide adaptation aid to those countries to which they also give other forms of development aid. This is not too surprising, given that adaptation is a subset of development aid and that there is no separate decision-making process for the former (see also Chap. 6). The strong link between adaptation aid and development aid is particularly evident in our quantitative models for the three selected country cases, Germany, Sweden, and the UK. We do find some evidence in these models that one or the other hypothesised drivers of adaptation aid allocation—recipient need, recipient merit, and donor interests—plays a role in all three countries, and more so at the selection than at the allocation stage. Yet, the results are comparatively weak—except for total development aid. Across the three case study countries, development aid is a very strong predictor of adaptation aid allocation. The qualitative results point in a similar direction. Interviewees emphasise that adaptation aid is part of development aid, and that past working relations are important for both forms of aid. Successful completion of development projects signals to donors that the recipient is capable of putting resources to good use and that development assistance is likely to be used effectively and efficiently. Furthermore, recipient merit and economic interests are also relatively strong drivers of adaptation aid allocation (especially at the selection stage, see panel (a) of Fig. 7.1).

Understandable as this is, these quantitative and qualitative findings imply that—indirectly—the allocation logic of development aid is transferred to adaptation aid. This also locks in allocation patterns that may be problematic—from the perspective of delivering adaptation aid to those most in need—such as path dependencies that lead to preferential treatment of recipients with long-lasting relationships to donors (see Barrett 2014; Robertsen et al. 2015) and network effects that lead to aid 'darlings' and 'orphans' (see e.g. Davies and Klasen 2013; Hoeffler and Outram 2011). Such patterns are not in line with promises made during the UNFCCC negotiations. So far, much of the criticism targeted at adaptation funding is connected to questions of additionality: adaptation finance should be over and above development aid and therefore the two aid flows must be more clearly separated from each other (e.g. Brown et al. 2010; Duus-Otterström 2015; Huhtala et al. 2010; Stadelmann et al. 2010; Weikmans 2016). However, the idea that development and climate aid (including of course adaptation aid) should be more clearly separated has also this second, less discussed dimension: that donors simply provide adaptation aid to the same set of recipient countries to which they provide development assistance. If this is the case, then the most important *raison d'être* for adaptation aid must be called into question. If donors simply provide adaptation aid to the same set of recipients, to what extent do they indeed support 'particularly vulnerable' countries in meeting the cost of climate change impacts as agreed to in the 1992 Convention? This relates back to the discussion of adaptation aid versus adaptation finance (see Chap. 2).

Adaptation finance—which is the subject of the UNFCCC negotiations—must be new and additional and prioritised on particularly vulnerable countries. Donors use their development aid budgets to comply with their climate finance commitments, and there are good reasons to do this. Yet, if adaptation aid is used to meet climate finance commitments, donors must not treat adaptation aid as development aid, but as climate finance. One of our interview partners aptly summarises this discussion:

> We see development assistance as charity that the rich countries have promised. They have not kept their promise, but that does not matter. They choose who to give it to, we cannot tell them, logically. Whereas, when we agreed under the UNFCCC treaty on funding to be given for adaptation, the treaty it is saying that the Annex I countries have obligations. When we agree to something under the UNFCCC, it is a treaty obligation. The UNFCCC is not a development treaty, it is a climate change treaty. The Annex I countries are the polluters, they have caused the problem. The developing countries are the victims of pollution. If they are being given money to help them deal with that pollution, it is a very different paradigm (UK2).

Because adaptation aid is not a gift, but an obligation, adaptation aid allocation should first and foremost be based on needs. However, our research shows that recipient need—although not insignificant—is only one among several determinants of adaptation aid allocation, and not the strongest one. Perhaps donor countries are still in a learning process of how to handle a new form of finance. So far, they rely to a large extent on their experience with development aid, but they have already started to diversify the allocation of adaptation aid and to consider both the physical vulnerability and adaptive capacity of recipients, albeit so far not to a desirable degree.

From a policy point of view, developing (non-Annex I) countries should closely monitor how donors distribute adaptation aid and insist on allocation based on need during the UNFCCC negotiations. From a research point of view, the observed allocation patterns represent an opportunity to study further how adaptation aid allocation *changes* over time, which we did not do in this study. Our study, however, provides a baseline for examining changing patterns over time, and as such, a first step towards a better understanding of adaptation aid and adaptation finance.

Note

1. Japan disbursed $5.7 billion (41.9% of its commitments) for adaptation projects between 2010 and 2015.

References

AdaptationWatch. (Weikmans et al.) 2016. Towards Transparency. The 2016 Adaptation Finance Transparency Gap Report. AdaptationWatch. White Paper. Available online at http://www.adaptationwatch.org/.

Ayers, J.M., and A.C. Abeysinghe. 2013. International Aid and Adaptation to Climate Change. In *The Handbook of Global Climate and Environmental Policy*, Chap. 28, ed. R. Falkner, 486–506. New York: Wiley.

Barrett, S. 2014. Subnational Climate Justice? Adaptation Finance Distribution and Climate Vulnerability. *World Development* 58: 130–142.

Bird, N. 2011, August. Going Beyond Aid Effectiveness to Guide the Delivery of Climate Finance. ODI Background Note. Overseas Development Institute, London.

Brown, J., N. Bird, and L. Schalatek. 2010, June. Climate Finance Additionality: Emerging Definitions and Their Implications. Climate Finance Policy Brief No.2. Heinrich Böll Stiftung. The Green Political Foundation. Available online at https://www.odi.org/sites/odi.org.uk/files/odi-assets/publications-opinion-files/6032.pdf.

Carty, T., J. Kowalzig, and A. Peterson. 2016, November. Climate Finance Shadow Report 2016: Lifting the Lid on Progress Towards the $100 Billion Commitment. Oxfam International. Available online at http://www.oxfamnovib.nl/Redactie/Pdf/bp-climate-finance-shadow-report-031116-en.pdf.

Ciplet, D., J.T. Robert, and M. Khan. 2013. The Politics of International Climate Adaptation Funding: Justice and Divisions in the Greenhouse. *Global Environmental Politics* 13(1): 49–68.

Davies, R.B., and S. Klasen. 2013. Of Donor Coordination, Free-Riding, Darlings, and Orphans: The Dependence of Bilateral Aid Commitments on Other Bilateral Giving. CESifo Working Paper No. 4177.

Donner, S.D., M. Kandlikar, and S. Webber. 2016. Measuring and tracking the flow of climate change adaptation aid to the developing world. *Environmental Research Letters* 11(5): 1–9.

Duus-Otterström, G. 2015. Allocating Climate Adaptation Finance: Examining Three Ethical Arguments for Recipient Control. *International Environmental Agreements* 16(6): 655–670.

Ha, S., T. Hale, and P. Ogden. 2016. Climate Finance in and between Developing Countries: An Emerging Opportunity to Build On. *Global Policy* 7(1): 102–108.

Hoeffler, A., and V. Outram. 2011. Need, Merit, or Self-Interest—What Determines the Allocation of Aid? *Review of Development Economics* 15(2): 237–250.

Huhtala, A., S. Curto, and P. Ambrosi. 2010, May. Monitoring Climate Finance and ODA. World Bank Issues Brief 1.

Junghans, L., and S. Harmeling. 2012. Different Tales from Different Countries: A First Assessment of the OECD 'Adaptation Marker'. Germanwatch Briefing Paper.

Klein, R.J. 2009. Identifying Countries that are Particularly Vulnerable to the Adverse Effects of Climate Change: An Academic or a Political Challenge? *Carbon & Climate Law Review* 3(3): 284–291.

Ministry for Foreign Affairs. 2007, August. Guidelines: Decisive Factors for Country Focus. Memorandum of the Department for Development Policy of the Ministry for Foreign Affairs, Stockholm.

OECD. 2013. oecd *Development Co-operation Peer Review: Sweden 2013.* Paris: OECD Publishing.

OECD. 2016. OECD.Stat Extracts. Available online at http://stats.oecd.org/.

Peterson Carvalho, A., and P. Terpstra. 2015. Tracking Adaptation Finance: An Approach for Civil Society Organizations to Improve Accountability for Climate Change Adaptation. Oxfam America Inc. and World Resources Institute. Available online at http://www.wri.org/publication/tracking-adaptation-finance.

Roberts, T.J., and R. Weikmans. 2015, December. The Unfinished Agenda of the Paris Climate Talks: Finance to the Global South. Brookings Series: Understanding COP21 and Beyond, Number 24. December 22. Available at: http://www.brookings.edu/blogs/planetpolicy/posts/2015/12/22-cop21-unfinished-agenda-finance-global-south-roberts-weikmans.

Roberts, J.T., and R. Weikmans. 2017. Postface: Fragmentation, Failing Trust and Enduring Tensions over What Counts as Climate Finance. *International Environmental Agreements: Politics, Law and Economics* 17(1): 129–137.

Roberts, T.J., R. Weikmans, and C. Jones. 2017. Allocating Climate Adaptation Finance: Shifting Positions in the Scholarly and Policy Realms.

Robertsen, J., N. Francken, and N. Molenaers. 2015. Determinants of the Flow of Bilateral Adaptation-Related Climate Change Financing to Sub-Saharan African Countries. LICOS Discussion Paper 373/2015. Catholic University Leuven.

Smith, J.B., T. Dickinson, J.D.B. Donahue, I. Burton, E. Haites, R.J. Klein, and A. Patwardhan. 2011. Development and Climate Change Adaptation Funding: Coordination and Integration. *Climate Policy* 11: 987–1000.

Stadelmann, M., J.T. Robert, and A. Michaelowa. 2010, November. Keeping a Big Promise: Options for Baselines to Assess "New and Additional" Climate Finance. Centre for Comparative and International Studies (ETH Zurich and University of Zurich) Working Paper Nr. 66.

UNFCCC. 1992. United Nations Framework Convention on Climate Change. Document FCCC/INFORMAL/84.

UNFCCC. 2007. Report of the Conference of the Parties on its thirteenth session, held in Bali from 3 to 15 December 2007 Addendum Part Two: Action taken by the Conference of the Parties at its thirteenth session. Document FCCC/CP/2007/6/Add.1.

UNFCCC. 2009. Report of the Conference of the Parties on its fifteenth session, held in Copenhagen from 7 to 19 December 2009. Addendum. Part Two: Action taken by the Conference of the Parties at its fifteenth session. Document FCCC/CP/2009/11/Add.1.

UNFCCC. 2015, December. Adoption of the Paris Agreement. Decision 1/CP.21. Document FCCC/CP/2015/10/Add.1.

Weikmans, R. 2016. Le rôle de la coopération au développement dans le financement international de l'adaptation au changement climatique. In *La nouvelle géographie du développement. Coopérer dans un monde en mutation*, ed. A. Zacharie, 175–185. Brussels: Éditions Le Bord de l'Eau.

Appendix

Table A.1 Summary statistics for all variables used in the quantitative analysis

	Mean	St. dev.	Median	Min.	Max.	Valid *N*
Principal adapt. aid[a]	0.025	0.216	0	0	7.432	22,962
Princ. and sign. adapt. aid[a]	0.051	0.291	0	0	7.432	22,962
ND-GAIN exposure	0.503	0.065	0.499	0.360	0.743	22,908
EVI	3.278	0.764	3.230	1.670	5.130	22,742
CRI (inverted)	51.427	31.908	47.500	0.660	123	20,946
GDP per capita[a]	7.819	1.061	8.025	5.369	10.139	22,380
WGI-index	−0.474	0.630	−0.472	−2.410	1.214	22,824
ND-GAIN adaptive capacity	0.563	0.161	0.536	0.283	0.944	22,908
Exports[a]	14.976	5.535	16.303	0	26.204	23,406
Distance[a]	8.618	0.821	8.795	0	9.848	22,908
UN voting	0.480	0.337	0.520	−1	1	22,880
Total aid[a]	1.093	1.567	0.183	0	8.732	23,406
Population[a]	15.560	2.251	15.909	9.191	21.034	22,990
African dummy	0.362	0.481	0	0	1	23,406
SIDS	0.248	0.432	0	0	1	23,406
LDCs	0.343	0.475	0	0	1	23,406
Colony dummy	0.029	0.168	0	0	1	23,406

[a]Natural logarithm used

C. Betzold and F. Weiler, *Development Aid and Adaptation to Climate Change in Developing Countries*,
https://doi.org/10.1007/978-3-319-64510-0

Table A.2 Questions guiding our semi-structured interviews

1. What is the role of [ORGANISATION/UNIT] in adaptation and in adaptation aid? How would you describe your personal role?
2. How do you see the role of climate change generally and climate change adaptation in particular in [COUNTRY'S] development cooperation?
    (a) How has climate change affected the way [ORGANISATION/UNIT] works (if at all)?
    (b) How does [COUNTRY] compare to other European and global donors with regard to climate change and aid?
    (c) How does adaptation compare to mitigation?
3. Can you help us understand the decision-making process on how much climate change and adaptation aid [COUNTRY] provides and to whom?
    (a) Countries receive different levels of support for adaptation from [COUNTRY]. Can you highlight some of the specific reasons why recipient countries receive more or less adaptation aid?
    (b) How do you evaluate current reporting on adaptation, especially in the OECD CRS using the Rio marker for adaptation?
4. How satisfied are you with the outcomes—that is the volume and allocation of adaptation aid by [COUNTRY]?
5. How do you see the future? Where is [COUNTRY'S] and global adaptation aid heading to after Paris?
6. Is there anything else we should know about climate change, adaptation, and aid?

Index

Symbol
$100 billion target, 9–10, 27, 28, 31, 39, 188, 202
0.7% target, 29, 59, 173, 192, 194

A
Absorptive capacity, 138, 142, 184, 195
Adaptation deficit, 4, 8
Adaptation finance, 58, 192, 212–214
Adaptation Fund, 10, 43, 105, 194
Adaptation marker, *see* Rio marker for adaptation
Adaptive capacity, 3–7, 30, 38, 40, 63, 67, 68, 72–76, 81, 122, 125, 137–149, 162, 166, 182, 203, 206–208, 211–213
African countries, 72–90, 138, 142–145, 162, 164, 165, 181, 184, 210
Aid effectiveness/efficiency, 39, 73, 74, 142, 179, 188, 207, 212
Aid Policy Framework, 86, 187

B
Bündnis 90/Die Grünen, 85
Bali Action Plan, 26
Bilateral Aid Review, 179
BMZ, 85
Bundesministerium für wirtschaftliche Zusammenarbeit und Entwicklung, *see* BMZ

C
Cameron, David, 32, 136
Cancùn COP, 27, 29
Care International, 86
CDKN, 86
CDU, 85
Charter for the Future, 183
Church of Sweden international, 85
Clean Development Mechanism, 10, 24, 25
Climate commitment, 88, 185
Climate finance, 2, 8, 10–11, 21, 28, 29, 65, 172, 191, 213
Climate Investment Funds, 105

C. Betzold and F. Weiler, *Development Aid and Adaptation to Climate Change in Developing Countries*,
https://doi.org/10.1007/978-3-319-64510-0

Climate justice, 9, 23, 178
Climate Risk Index, *see* CRI
Climate variability, 4, 70, 135, 176
Climate-proof, 8, 174, 176, 177, 191, 192
Common but differentiated responsibilities and respective capabilities, 23
Convention, *see* UNFCCC
Copenhagen Accord, 22, 26–29, 178, 191, 202, 210
Cost of adaptation, 7–9
Creditor Reporting System, *see* OECD CRS
CRI, 58, 67, 70–71, 81, 83, 130, 135, 136, 162, 163, 165, 207
Cshapes package, 77, 152

D
Department for International Development, *see* DfID
Deutsche Gesellschaft fr International Zusammenarbeit, *see* GIZ
Deutsches Institute für Entwicklungspolitik, *see* DIE
DEval, 85
Development Assistance Committee, *see* OECD DAC
DfID, 86, 174
Diakonia, 85
DIE, 85
Donor interests, 12, 22, 32–35, 43, 46, 57, 61, 66, 76–79, 81, 86–89, 121, 122, 125, 149–155, 162, 164, 166, 172, 186–187, 201, 203, 206, 207, 209, 212
Durban COP, 27

E
Earth Summit, 22, 62
Energy and Climate Fund, 174
Environmental Vulnerability Index, *see* 217
Equity, *see* Climate justice
European Bank for Reconstruction, 105
European Union, 104
EVI, 58, 67, 69–70, 81, 83, 130, 134–136, 162, 163, 165, 207
Exposure and sensitivity, 3–7, 38, 40, 67–69, 73, 75, 122, 130–137, 171, 181, 203
Extreme weather events, 1, 43, 135

F
Fast-start finance, 9, 11, 27, 42, 174, 202
Fondation pour les Etudes et Recherches sur le Développement International, 70

G
GEF, 23, 105
Germanwatch, 70–71, 85, 135
Germany, 3, 12, 14, 57–60, 79, 80, 82–86, 89, 100, 103, 106–108, 111, 117, 123, 156–163, 167, 201, 204, 205, 208–209, 212
GIZ, 60, 85
Global Environmental Facility, *see* GEF
Good governance, 35, 38, 43, 72–76, 88, 122, 137–149, 162, 164, 166, 183–185, 203, 207, 208, 212
Green Climate Fund, 27, 43, 105, 190, 194

I
Intergovernmental Panel on Climate Change, *see* IPCC
International Climate Fund, 175

International Climate Initiative, 174
International Development (Official Development Assistance Target) Act 2015, 29, 173, 194
International Fund for Agricultural Development, 105
IPCC, 1, 4–6

K
KfW, 60, 85
Kyoto Protocol, 22–26, 43

L
Landscape of Climate Finance, 10, 196
LDCs, 72–90, 124, 138, 142–145, 162, 164, 165, 181, 184, 189, 210
Least Developed Country Fund, 25, 105
Least Developing Countries, *see* LDCs
Loss and damage, 7, 27

M
Mainstreaming, 105, 174, 175
Marrakesh Accords, 22, 25
Merkel, Angela, 194
Mitigation, 4, 9–11, 23, 25, 62, 101, 177–178, 202
Monterrey Consensus, 35
Multilateral aid, 10, 31, 90, 100, 104–105, 109, 111–113, 178, 189, 205

N
Naturskyddsföreningen, 85
ND-GAIN, 42, 58, 67–69, 75–76, 81, 83, 125, 130–133, 136, 137, 145–148, 162–165, 207, 212
New and additional/additionality, 10, 14, 23, 28–31, 89, 167, 172, 176, 190–192, 209–214
Nordic Development Fund, 105
Notre Dame Global Adaptation Index, *see* ND-GAIN

O
ODA, 2–4, 30–31, 44, 59, 73, 99–102, 105–110, 117, 122–123, 142, 149, 152, 155–156, 165, 167, 191, 202, 205, 209, 211–214
ODI, 86
OECD, 3, 4, 13, 23, 39, 63, 103, 105, 133
  OECD CRS, 14, 46, 60, 62–65, 80, 99–101, 104, 108, 109, 117, 122, 133, 192, 194, 204
  OECD DAC, 3, 31, 57, 62, 80
Official development aid, *see* ODA
Over-reporting, 64, 192, 193, 211
Overseas Development Institute, *see* ODI
Ownership, 182, 185
Oxfam, 8, 86
Oxfam Germany, 85

P
Paris Agreement, 2, 12, 22, 27–28, 39, 178, 191, 195, 210
Particularly vulnerable countries, 9, 24, 38–39, 43, 72–73, 81, 142–145, 181–182, 202, 210, 211
Physical vulnerability, 58, 66–71, 73, 81, 122, 125, 130–137, 143, 149, 157, 163, 165, 166, 203, 206–208, 211, 213
Policy for Global Development, 183
Private finance, 10, 172, 188–189

Q
Quality of Government Institute Dataset, 78, 153

R
Recipient merit, 12, 22, 32, 35–36, 43, 46, 57, 61, 66, 72–76, 81, 86–89, 121, 125, 137–149, 164, 166, 172, 183–185, 201, 206–208, 212
Recipient need, 12, 22, 32–35, 38, 46, 57, 61, 66–71, 75, 81, 86–89, 121, 130–149, 162, 163, 165, 166, 172, 180–182, 203, 207, 208, 212, 213
Reporting and accounting, 89, 172, 176, 192–194
Resilience, 6, 63, 175–177, 210
Rio marker for adaptation, 13, 62–63, 99, 100, 103, 177, 193, 194, 204

S
Sensitivity, *see* Exposure and sensitivity
Sida, 60, 85, 86
SIDS, 13, 42, 72–90, 113, 116, 117, 125, 138, 142–145, 162, 163, 165, 171, 181, 206, 210
Small country bias, 37, 44, 79, 122, 156, 166
Small Islands Developing States, *see* SIDS
SOPAC, 69
South Pacific Applied Geoscience Commission, *see* SOPAC
South-South climate finance, 11
Special Climate Change Fund, 25, 105
Special Climate Change Initiative, 174
Stern review, 8
Stockholm Environment Institute, 85
Structural Vulnerability to Climate Change Index, *see* SVCCI
SVCCI, 70
Sweden, 3, 12, 14, 57–60, 80, 82–86, 88, 89, 100, 103, 107–109, 111, 117, 123, 156, 163–165, 167, 201, 204, 205, 208–209, 212
Swedish International Development Cooperation Agency, *see* Sida
Swedish Ministry for Foreign Affairs, 85
Swedish Ministry of the Environment and Energy, 85

U
UN Comtrade Database, 76, 83, 150
UN General Assembly, 43, 78, 81, 149, 154, 163, 164, 207
UN General Assembly Voting Data, 78, 83, 154
UNDP, 8
UNFCCC, 2, 12, 22–25, 39, 88, 155, 181, 202, 210–214
  UNFCCC Standing Committee on Finance, 11, 27
United Kingdom, 3, 12, 14, 57–60, 80, 82–86, 89, 100, 103, 107, 108, 111, 117, 123, 156, 165–167, 201, 204, 205, 208–209, 212
United Nations, 76, 154
United Nations Development Programme, *see* UNDP
United Nations Framework Convention on Climate Change, *see* UNFCCC

V
Visibility, 172, 189, 194
Vulnerability, 3, 4, 38, 40–41, 171, 180–182, 205

See also Adaptive capacity; Physical vulnerability, 217

W
We Effect, 85
WGI, 74–75, 81, 83, 125, 137, 148, 147–212
World Bank, 8, 73, 74, 90
Worldwide Governance Indicators, *see* WGI

The manufacturer's authorised representative in the EU is Springer Nature Customer Service Centre GmbH, Europaplatz 3, 69115 Heidelberg, Germany. If you have any concerns regarding our products, please contact ProductSafety@springernature.com

Printed and bound by CPI Group (UK) Ltd, Croydon, CR0 4YY
25/09/2026
02222890-0004